Lecture Notes in Physics

New Series m: Monographs

The Editorial Policy for Monographs

The series Lecture Notes in Physics reports new developments in physical research and teaching - quickly, informally, and at a high level. The type of material considered for publication in the New Series m includes monographs presenting original research or new angles in a classical field. The timeliness of a manuscript is more important than its form, which may be preliminary or tentative. Manuscripts should be reasonably self-contained. They will often present not only results of the author(s) but also related work by other people and will provide sufficient motivation, examples, and applications.

The manuscripts or a detailed description thereof should be submitted either to one of the series editors or to the managing editor. The proposal is then carefully refereed. A final decision concerning publication can often only be made on the basis of the complete manuscript, but otherwise the editors will try to make a preliminary decision as definite as they can on the basis of the available information.

Manuscripts should be no less than 100 and preferably no more than 400 pages in length. Final manuscripts should preferably be in English, or possibly in French or German. They should include a table of contents and an informative introduction accessible also to readers not particularly familiar with the topic treated. Authors are free to use the material in other publications. However, if extensive use is made elsewhere, the publisher should be informed.

Authors receive jointly 50 complimentary copies of their book. They are entitled to purchase further copies of their book at a reduced rate. As a rule no reprints of individual contributions can be supplied. No royalty is paid on Lecture Notes in Physics volumes. Commitment to publish is made by letter of interest rather than by signing a formal contract. Springer-Verlag secures the copyright for each volume.

The Production Process

The books are hardbound, and quality paper appropriate to the needs of the author(s) is used. Publication time is about ten weeks. More than twenty years of experience guarantee authors the best possible service. To reach the goal of rapid publication at a low price the technique of photographic reproduction from a camera-ready manuscript was chosen. This process shifts the main responsibility for the technical quality considerably from the publisher to the author. We therefore urge all authors to observe very carefully our guidelines for the preparation of camera-ready manuscripts, which we will supply on request. This applies especially to the quality of figures and halftones submitted for publication. Figures should be submitted as originals or glossy prints, as very often Xerox copies are not suitable for reproduction. In addition, it might be useful to look at some of the volumes already published or, especially if some atypical text is planned, to write to the Physics Editorial Department of Springer-Verlag direct. This avoids mistakes and time-consuming correspondence during the production period.

As a special service, we offer free of charge LaTeX and TeX macro packages to format the text according to Springer-Verlag's quality requirements. We strongly recommend authors to make use of this offer, as the result will be a book of considerably improved technical quality. The typescript will be reduced in size (75% of the original). Therefore, for example, any writing within figures should not be smaller than 2.5 mm.

Manuscripts not meeting the technical standard of the series will have to be returned for improvement.

For further information please contact Springer-Verlag, Physics Editorial Department II, Tiergartenstrasse 17, W-6900 Heidelberg, FRG.

Jens Hoppe

Lectures on
Integrable Systems

Springer-Verlag
Berlin Heidelberg New York
London Paris Tokyo
Hong Kong Barcelona
Budapest

Author

Jens Hoppe
Institute for Theoretical Physics, Karlsruhe University
P. O. Box 6980, W-7500 Karlsruhe, Fed. Rep. of Germany

ISBN 3-540-55700-8 Springer-Verlag Berlin Heidelberg New York
ISBN 0-387-55700-8 Springer-Verlag New York Berlin Heidelberg

Typesetting: Camera ready by author/editor using the TeX macro package from
Springer-Verlag
Printing: Druckhaus Beltz, Hemsbach/Bergstr.
Bookbinding: Buchbinderei Kränkl, Heppenheim
58/3140-543210 - Printed on acid-free paper

Preface

These lectures were given at Karlsruhe University during the year 1991. They were aimed at graduate students working for their Diplom (which meant that roughly one third was specializing in one of the fields solid state physics, field theory or high energy physics). Though a more general elegant description using the language of differential geometry and the structure theory of semi-simple Lie algebras (among others) would have been further reaching and, perhaps, more appealing, I believe that (at least for a one semester course) simple case-studies involving rather elementary step by step calculations, while also staying closer to the original development, are usually more helpful to someone beginning his or her own research. Hence I have been as explicit as possible.

In writing up the material, I hope that this simple presentation may also be useful to someone who is more advanced, but hesitating to go through too many generalities before getting to a concrete problem. Integrable systems, following a rather dormant century, and one or two decades of quiet work (mostly done and appreciated in the Soviet Union and Japan, and a small number of people from the West) have recently gained a broader audience, and will very likely be of key interest again (as they were when the analytical description of dynamical systems was put on solid ground). In any case, knowing exact properties of systems with non-trivial dynamics should be useful in many different fields of physics.

Most of what is covered here has been known for some time, and part of what is new is based on collaborations with Martin Bordemann, Mikhail Olshanetsky, Peter Schaller, Martin Schlichenmaier and Stefan Theisen. Invaluable help in typesetting the manuscript was provided by Peter Maurer and Jürgen Schulze. To all of them I would like to express my thanks.

Karlsruhe, January 1992   Jens Hoppe

Contents

1 The Projection Method of Olshanetsky and Perelomov

Let us consider a finite dimensional system of point particles described by a classical Hamiltonian of the form $H(\mathbf{p}, \mathbf{q}) = \frac{1}{2}\mathbf{p}^2 + V(\mathbf{q})$. Ideally, one would like to know $\mathbf{q}(t) = (q_1(t), \ldots, q_N(t))$ as solution to the (usually coupled, nonlinear) ordinary differential equations $\ddot{\mathbf{q}} = -\nabla V(\mathbf{q}(t))$, for given $V(\mathbf{q})$ and initial conditions $\mathbf{q}(0) = \mathbf{a}$, $\dot{\mathbf{q}}(0) = \mathbf{p}(0) = \mathbf{b}$. Unfortunately, it is by far the *exceptional* situation to be able to determine $\mathbf{q}(t)$ analytically. Moreover, according to the degree of practicality in the definition of 'knowing', one should distinguish between the possibility of explicitly writing down $\mathbf{q}(t)$ in terms of known functions from cases where the solutions can be obtained in terms of (finitely many) algebraic operations, the evaluation of integrals of known functions, their inversion, ... Already in the nineteenth century the latter possibility (sometimes referred to as 'solution by quadrature') was found to be implied by the existence of sufficiently many independent, (with H and among themselves) Poisson-commuting functions of $\mathbf{q}$ and $\mathbf{p}$ ('Liouville-integrability'); however, rarely allowing closed form solutions *in practice*.

In order to construct non-trivial examples in which the solutions to the equations of motion *can* be given (directly), one may try to find systems, whose (complicated) dynamics is merely the result of a 'non-linear' projection onto part of some 'freely' developing bigger system. To illustrate this possibility, let $X(t)$ be a hermitean $N \times N$ matrix. Write

$$X(t) = U(t)Q(t)U^{-1}(t) \quad, \quad Q(t) = \begin{pmatrix} q_1(t) & & & & 0 \\ & q_2(t) & & & \\ & & q_3(t) & & \\ & & & \ddots & \\ 0 & & & & q_N(t) \end{pmatrix} \tag{1.1}$$

i.e. diagonalize X for all t. In this case, the projetion method [1] consists of prescribing some simple dynamics for $X(t)$, e.g.

$$\ddot{X}(t) = 0 \quad, \tag{1.2}$$

while investigating how $Q(t)$ then varies in time. The $q_i(t)$, to be interpreted as the positions of N particles moving on a line, will obviously be the eigenvalues of

$$X(t) = X(0) + \dot{X}(0)\, t \quad. \tag{1.3}$$

On the other hand, (1.1) and (1.2) also imply the dynamics of Q via

$$\dot{X}(t) = U(t)L(t)U^{-1}(t) \quad \text{with} \quad L := [U^{-1}\dot{U}, Q] + \dot{Q} \quad ,$$

$$\ddot{X}(t) = U(t)\left([U^{-1}\dot{U}, L] + \dot{L}\right)U^{-1}(t) \overset{!}{=} 0$$

$$(1.4)$$

i.e.

$$\dot{L} = [L, M] \quad , \qquad M = U^{-1}\dot{U} \quad . \tag{1.5}$$

Note that $L^\dagger = L$, $M^\dagger = -M$ (as $X^\dagger = X$). Also note that (1.2) implies

$$C := [X(t), \dot{X}(t)] = \text{const} \quad ; \qquad C^\dagger = -C \quad . \tag{1.6}$$

Writing out (1.5), and seperating its diagonal from its off-diagonal part, yields

$$\ddot{q}_i(t) = -\sum_{k \neq i} 2M_{ik}M_{ki}(q_i - q_k) \quad , \tag{1.7}$$

$$\dot{M}_{ij}(q_i - q_j) + 2M_{ij}(\dot{q}_i - \dot{q}_j) = \sum_k M_{ik}M_{kj}(q_i + q_j - 2q_k) \quad (i \neq j) \quad . \tag{1.7a}$$

For given initial conditions $M(t)$ is of course determined via $M = U^{-1}\dot{U}$, i.e. by obtaining U from $X(0) + \dot{X}(0)t$ (prescribing some order for the eigenvalues). This being inpractable for general N, however, one must try to find $M(t)$ by more indirect arguments (and special choices of C):
Suppose we started with $X(0)$ being diagonal, i.e.

$$U(0) = \mathbb{1} \tag{1.8}$$

(leaving as initial conditions $Q(0)$ and $L(0) = \dot{X}(0)$).
Then $C_{ij} = [Q(0), L(0)]_{ij}$, hence

$$L_{i \neq j}(0) = \frac{C_{ij}}{q_i(0) - q_j(0)} \quad . \tag{1.9}$$

As (for all t):

$$L_{ij}(t) = \delta_{ij}\dot{q}_j(t) - M_{ij}(t)(q_i(t) - q_j(t)) \quad , \tag{1.10}$$

one also knows that

$$L_{ii}(0) = \dot{q}_i(0) \quad , \qquad M_{i \neq j}(0) = \frac{-C_{ij}}{(q_i(0) - q_j(0))^2} \quad . \tag{1.11}$$

Taking this as motivation to assume

$$M_{i \neq j}(t) = -\frac{C_{ij}}{(q_i(t) - q_j(t))^2} \tag{1.12}$$

for *all* t, which via (1.10) implies

$$L_{ij}(t) = \delta_{ij}\dot{q}_j(t) + \frac{C_{ij}}{(q_i(t) - q_j(t))} \quad , \tag{1.13}$$

(1.7) becomes

$$\ddot{q}_i(t) = -2 \sum_{k \neq i} \frac{C_{ik} C_{ki}}{(q_i(t) - q_k(t))^3} \quad , \tag{1.14}$$

which are the classical equations of motion for a system of N particles moving on a line, whose 'energy' is

$$H = \frac{1}{2} \left(\sum_{i=1}^{N} p_i^2 - \sum_{i \neq j} \frac{C_{ij} C_{ji}}{(q_i - q_j)^2} \right) \quad , \tag{1.15}$$

while the remaining equations (1.7a) are

$$\frac{C_{ij}}{(q_i - q_j)} (M_{jj} - M_{ii}) + \sum_{k}{}'' \frac{C_{ik} C_{kj}}{(q_i - q_k)^2 (q_j - q_k)^2} (q_i + q_j - 2q_k) = 0 \quad (i \neq j) \tag{1.16}$$

(each dash in $\sum_{k}{}''$ indicating an obvious value of k to be omitted; i.e. $k \neq i$ and $k \neq j$). Note that the l.h.s. of (1.7a) has become identically zero, which could be taken as an independent justification for (1.12). While it is not clear whether these equations can be satisfied for general C_{ij} (for real antisymmetric C they certainly can *not* — which simply means that (1.12) does not hold in that case)

$$C_{jk} = ig(1 - \delta_{jk}) \quad , \qquad M_{jj} = ig \sum_{k}{}' \frac{1}{(q_j - q_k)^2} \tag{1.17}$$

solves (1.16), which means that the classical system of N point particles governed by

$$H = \frac{1}{2} \left(\sum_{i=1}^{N} p_i^2 + g^2 \sum_{i \neq j} \frac{1}{(q_i - q_j)^2} \right) \tag{1.15'}$$

is completely integrable, in the sense that for (almost) arbitrary initial conditions $(q_1 < q_2 < \ldots < q_N)$ the corresponding equations of motion can be (and are) solved by the eigenvalues of

$$X(t) = \begin{pmatrix} a_1 + b_1 t & & \frac{igt}{a_j - a_k} \\ & \ddots & \\ \frac{igt}{a_j - a_k} & & a_N + b_N t \end{pmatrix} \quad , \tag{1.18}$$

$$a_j = q_j(0) \quad , \quad b_j = p_j(0) = \dot{q}_j(0) \quad .$$

Leaving aside cases where the equations of motion are *directly* solved in terms of known functions, the above situation is about as 'good' as one can possibly hope.

What would have happened if we had not prescribed completely free dynamics for X, but some other simple evolution equation that allows an explicit solution for $X(t)$, like

$$\ddot{X}(t) + \omega^2 X(t) = 0 \quad , \tag{1.19}$$

$$X(t) = X(0) \cos \omega t + \frac{\dot{X}(0)}{\omega \sin \omega t} \quad . \tag{1.20}$$

We could still consider the diagonalization of X (Eq.(1.1)) and look for the dynamics of the eigenvalues of X as implied by (1.20). Instead of (1.5), we will get

$$\ddot{X} = U\left([M,L] + \dot{L}\right)U^{-1} \overset{!}{=} -\omega^2 X = -\omega^2 U Q U^{-1} \quad , \tag{1.21}$$

i.e.

$$\ddot{Q} + [\dot{M},Q] = \dot{L} = [L,M] - \omega^2 Q = [\dot{Q},M] + [[M,Q],Q] - \omega^2 Q \quad . \tag{1.22}$$

If we choose M (as a matrix–valued function of the $q_i(t)$) as before (cp. (1.12) / (1.17)) the first order derivative terms will again cancel, and one obtains

$$\ddot{Q} = [[M,Q],M] - \omega^2 Q \quad . \tag{1.23}$$

$[[M,Q],M]$ will be purely diagonal (due to (1.16)) so that (1.23) is consistent — provided the equations of motion

$$\ddot{q}_i(t) = 2g^2 \sum_k{}' \frac{1}{(q_i - q_k)^3} - \omega^2 q_i(t) \tag{1.24}$$

hold. This means that the eigenvalues $q_i(t)$ of $X(t)$ (given as in (1.20)) develop in time just as the positions of N particles on a line whose dynamics is governed by

$$H = \frac{1}{2}\left(\sum_{i=1}^{N} p_i{}^2 + g^2 \sum_{i \neq j} \frac{1}{(q_i - q_j)^2} + \omega^2 \sum_{i=1}^{N} q_i{}^2\right) \quad . \tag{1.25}$$

As is rather clear from the close similarity in the projections, and as we will also see in the algebraic treatment of the quantum mechanical case, systems of type (1.25) with $\omega \neq 0$ and $\omega = 0$, respectively, are strongly related. In the classical case, one can show (as an example to a much more general statement [3]) that if $q_i'(t)$ solves (1.24) with $\omega = 0$,

$$q_i(t) = q_i'\left(\frac{1}{\omega}\tan\omega t\right)\cos\omega t \tag{1.26}$$

solves (1.24) with $\omega \neq 0$.

Let us conclude by looking at yet another type of projection (which will lead to pair potentials proportional to $1/\sinh^2(ax)$) [1]: Consider a positive definite hermitean matrix X, with unit determinant. Due to the positivity of its eigenvalues it is useful to write its diagonalisation in the form

$$X(t) = U(t)e^{2aQ(t)}U^{-1}(t) \tag{1.27}$$

with

$$Q(t) = \text{diag}(q_1(t), \ldots, q_N(t)) \tag{1.28}$$

real. Differentiating (1.27) one gets

$$\dot{X}(t) = U(t)\left([M, e^{2aQ}] + 2a\dot{Q}e^{2aQ}\right)U^{-1}$$
$$M = U^{-1}(t)\dot{U}(t) \quad . \tag{1.29}$$

'Unfortunately' it is meaningless to consider $\ddot{X} = 0$ in this case as the free un-coupled development of the individual matrix elements will in general destroy the positive definiteness of X. What corresponds to a free motion in the space of positive definite matrices (of unit determinant) is the equation [1]

$$\frac{d}{dt}(X^{-1}\dot{X} + \dot{X}X^{-1}) = 0 \quad , \tag{1.30}$$

which is solved by

$$X(t) = Be^{2At}B^\dagger$$
$$B \in \mathrm{SL}(N,\mathbb{C}) \quad , \quad A^\dagger = A \quad , \quad \mathrm{Tr}\,A = 0 \quad . \tag{1.31}$$

Note that X is indeed positive definite, as it is of the form $C^\dagger C$. In analogy with $\dot{X} = ULU^{-1}$ (in the case $\ddot{X} = 0$) let us define $L(t)$ via

$$4a\,UL(t)U^{-1} = X^{-1}\dot{X} + \dot{X}X^{-1} \equiv J \quad . \tag{1.32}$$

So

$$L(t) = \dot{Q} + \frac{1}{4a}(e^{-2aQ}Me^{2aQ} - e^{2aQ}Me^{-2aQ}) \quad , \tag{1.33}$$
$$\dot{L} = [L, M] \quad , \quad M = U^{-1}\dot{U} \quad .$$

The factors of a (and $1/a$) are such that $L(t)$ (as in (1.33)) goes to $\dot{Q}+[M,Q]$ (as it was defined in (1.4)) when $a \to 0$. Again one can try to use the conserved quantities $(J,$ Eq. (1.32)) to find L (and M) in terms of the $q_i(t)$ and initial conditions. For the special case $B = e^{aQ(0)}$ ($X(0)$ diagonal, $U(0) = \mathbb{1}$) one finds [1]

$$L_{ij}(t) = \delta_{ij}\dot{q}_i(t) + iag\coth(a(q_i(t) - q_j(t)))(1 - \delta_{ij}) \tag{1.34}$$

as a consistent choice for L, and the equations of motion corresponding to

$$H := \frac{1}{2}\mathrm{Tr}\,L^2 - \frac{1}{2}\frac{N(N-1)}{2}a^2g^2 = \frac{1}{2}\left(\sum_{i=1}^{N}p_i^2 + \sum_{i\neq j}\frac{a^2}{\sinh^2(a(q_i - q_j))}\right) \tag{1.35}$$

to be fulfilled by $\frac{1}{2a}$ times the logarithm of the eigenvalues of

$$X(t) = e^{aQ(0)}e^{2At}e^{aQ(0)} \quad , \tag{1.36}$$

where

$$A_{jk} = a\delta_{jk}\dot{q}_j(0) + ia^2(1 - \delta_{jk})\frac{1}{\sinh a(q_j - q_k)} \quad . \tag{1.37}$$

Notes and References

Analogous methods of projection exist for the corresponding quantum systems [2]. For the geometric meaning in both cases see [1], [2].

[1]   M.A. Olshanetsky, A.M. Perelomov; Lett. Nuov. Cim. 16 (1976) 333.
                                        Lett. Nuov. Cim. 17 (1976) 97.
                                        Phys. Rep. 71 (1981) 314.
[2]   M.A. Olshanetsky, A.M. Perelomov; Phys. Rep. 94 #6 (1983) 313.
[3]   A.M. Perelomov; Comm. Math. Phys. 63 (1978) 9.

2 Classical Integrability of the Calogero-Moser Systems

We have seen that if one defines an $N \times N$ matrix L as

$$(L)_{jk} = \delta_{jk} p_j + ig \frac{(1 - \delta_{jk})}{q_j - q_k} \tag{2.1}$$

one can write the equations of motion belonging to

$$H = \frac{1}{2} \left(\sum_{i=1}^{N} p_i^2 + g^2 \sum_{i \neq j} \frac{1}{(q_i - q_j)^2} \right) \tag{2.2}$$

in the form

$$\dot{L} = [L, M] \quad , \tag{2.3}$$

where M is given by

$$M_{jk} = ig\delta_{jk} \sum_{l \neq j} \frac{1}{(q_j - q_l)^2} - ig \frac{(1 - \delta_{jk})}{(q_j - q_k)^2} \quad . \tag{2.4}$$

The crucial observation, made by P. Lax [1] (in the context of nonlinear partial differential equations, i.e. differential operators, instead of finite dimensional matrices) is that any equation of type (2.3), independent of the nature of the specific form of L and M, implies the existence of infinitely many time-independent quantities,

$$C_k := \operatorname{Tr} L^k \quad , \quad \frac{d}{dt} C_k = 0 \quad , \quad (k = 1, 2, \ldots) \tag{2.5}$$

– as long as there exists an invariant trace , $\operatorname{Tr}(AB) = \operatorname{Tr}(BA)$. In the case of L being an $N \times N$ Matrix, with the trace being defined as the sum of all diagonal elements, only N of the C_k will be algebraically independent, of course (as the C_k are invariant under $L \to ULU^{-1}$; so they are functions of the N eigenvalues of L – of which there are at most N independent ones (equivalently, one may consider $\det(L - \lambda \mathbb{1})$, written as an N-th order polynomial in λ, replace λ by L and take the trace – resulting in an explicit expression of $\operatorname{Tr} L^N$ as a polynomial in the $\operatorname{Tr} L^k$, $k < N$, and $\det L$; the same holds for all higher C_k, $k > N$). The existence of the constants of motion (2.5) certainly simplifies the dynamics. Moreover, *if* the C_k Poisson-commute, one can in principle(!) take them as new momenta

$$P_k := C_k \qquad (k = 1, 2, \ldots, N) \quad , \tag{2.6}$$

with new coordinates Q_k chosen such that the transformation to Q and P variables is canonical; the equations of motion (written in these new variables) then become trivial:

$$\dot{P}_k = 0 \quad , \quad \dot{Q}_k = \frac{\partial H}{\partial P_k} \quad . \tag{2.7}$$

In 'most cases' (as for (2)) , $C_2 = 2H$ (up to numerical constants), so that the solution of (7) is

$$Q_2 = \frac{1}{2}t + d_2 \quad Q_{k \neq 2} = d_k \quad (P_k = \text{const., all } k) \quad , \tag{2.8}$$

with d_k being some numerical constants.

In any case, let us try to find other dynamical systems for which the equations of motion can be written in the (Lax) form (2.3).

One possibility [2],[3] is to note that (2.1) is of the form

$$L = \Pi + igF \tag{2.9}$$

with $\Pi = \text{diag}(p_1, p_2, \ldots, p_N)$ being diagonal, and F,

$$F_{j \neq k} = f(q_j - q_k) \tag{2.10}$$

being purely off-diagonal, and having its jk–component depend only on $q_j - q_k$. In (2.1),

$$f(x) = \frac{1}{x} \quad , \tag{2.11}$$

but one may try other possibilities. Letting f unspecified, and taking

$$M = ig(Z + Y) \tag{2.12}$$

with Z diagonal, and Y purely off-diagonal (and both independent of the momenta), one finds that (2.3) is equivalent to

$$\dot{F} = [\Pi, Y] \tag{2.13a}$$

$$\dot{\Pi} = -g^2[F, Z + Y] \quad . \tag{2.13b}$$

While (2.13a) is satisfied once

$$Y_{j \neq k} = f'(q_j - q_k) \quad , \tag{2.14}$$

(2.13b) has to be seperated into its diagonal part,

$$\dot{p}_i = -g^2[F, Y]_{ii} \tag{2.15}$$

(determining the potential in terms of f; remember that the r.h.s. is to equal $-\frac{\partial V}{\partial q_i}$), and its off-diagonal content,

$$[F, Z + Y]_{i \neq j} = 0 \quad ,$$

i.e.

$$f(q_i - q_j)(z_j - z_i) + \sum_k {}'' \left[f(q_i - q_k)f'(q_k - q_j) - \right.$$

$$\left. f'(q_i - q_k)f(q_k - q_j) \right] = 0 \quad , \qquad (2.16)$$

which has to hold for all $i \neq j$.

Making the additional Ansatz

$$z_i = \sum_{k \neq i} {}' z(q_i - q_k) \qquad (2.17)$$

one finds, that it is actually consistent to require *each* term in the sum over k to be zero; putting, for any such term,

$$x = q_i - q_k \qquad , \qquad y = q_k - q_j$$

one finds that (2.16) will be satisfied (for all $i \neq j$) if f and z satisfy the functional equation

$$f(x)f'(y) - f'(x)f(y) = (z(x) - z(-y))\, f(x+y) \qquad (2.18)$$

As a check one may verify that $f(x) = 1/x$, $z(x) = 1/x^2$ indeed satify (18). As found in [9],[10] and [11], (18) implies

$$z(x) = \frac{f''(x)}{2f(x)} \quad , \qquad (2.19)$$

and the solutions for $f(x)$ are such that $v(x) = -f(x)f(-x)$ is of one of the following types:

$$\frac{1}{x^2} \quad \text{(I)} \qquad\qquad \frac{a^2}{\sinh^2 ax} \quad \text{(II)}$$

$$\frac{a^2}{\sin^2 ax} \quad \text{(III)} \qquad\qquad a^2 \wp(ax) \quad \text{(IV)} \qquad (2.20)$$

where $\wp$ denotes the elliptic Weierstrass function (see e.g. [4]). Due to (14), the corresponding Hamiltonians are

$$H = \frac{1}{2} \sum_{i=1}^{N} p_i^2 + \frac{1}{2}g^2 \sum_{i \neq j} v(q_i - q_j) \quad . \qquad (2.21)$$

The potentials I, II and III are in some sense 'special cases' (limits) of IV, but it is useful to consider them seperately.

What remains to be shown is that there exist N independent Poisson commuting functions of the $2N$ variables $q_1, \ldots, p_N$. Due to (2.3), we know that all trace-polynomials of L ($=$ Traces of polynomials in L) Poisson commute with H. But do they commute among themselves? For systems of type I and II, this is actually trivially true, as the repulsive interaction implies that at $t = \pm\infty$ the particles behave as free particles, and that $C_k = \operatorname{Tr} L^k$ approaches $\sum_{i=1}^{N} p_i^k$ in both limits. As the C_k are time independent, their mutual commutativity and functional indepence (which is also obvious due to the different powers of the momenta) at a particular time ($t = \pm\infty$) suffices to imply these properties for all times. For systems with bounded motions, however, Poisson commutativity of the conserved charges is more difficult to prove. One possibility is to show that all eigenvalues of L Poisson commute (as $\operatorname{Tr} L^k = \sum_i \lambda_i{}^k$, one implies the other). Although it is in general impossible to obtain explicit expressions for the eigenvalues of L as functions of q and p, one can use *indirect* arguments [5], as is shown below, for the case of L given by

$$L_{jk} = \delta_{jk} p_j + i f(q_j - q_k) \tag{2.22}$$

– the coupling constant g is put here $= 1$, for simplicity –, and for the function f satisfying the functional equation (2.18) that led to the potentials of type I–IV (Eq. (2.20), and z given by (2.19)). The idea is to consider two different eigenvalues of L, λ and μ, with normalized (!) eigenvectors Φ and Ψ ($\in \mathbb{C}^N$)

$$
\begin{aligned}
(P + iF)\Phi = \lambda\Phi \qquad (\Phi, \Phi) = 1 \\
(P + iF)\Psi = \mu\Psi \qquad (\Psi, \Psi) = 1 \quad ,
\end{aligned}
\tag{2.23}
$$

and to show then that

$$\{\lambda, \mu\} := \sum_{j=1}^{N} \left(\frac{\partial\lambda}{\partial p_j} \frac{\partial\mu}{\partial q_j} - \frac{\partial\lambda}{\partial q_j} \frac{\partial\mu}{\partial p_j} \right) \equiv 0 \quad . \tag{2.24}$$

Due to (2.23), $\frac{\partial\lambda}{\partial q_k}$ and $\frac{\partial\lambda}{\partial p_k}$ (the same for μ) can be calculated in terms of f, f' and the components Φ_k, Ψ_k ($k = 1, \ldots, N$) of the two eigenvectors, Φ and Ψ:

$$(\Phi, \Phi) = \sum_{k=1}^{N} \Phi_k^* \Phi_k = 1$$

$$\Rightarrow \sum_{k=1}^{N} (\partial\Phi_k^*)\Phi_k + \Phi_k^*(\partial\Phi_k) = 0 \quad . \tag{2.25}$$

So (not always indicating sums over repeated indices)

$$
\begin{aligned}
\frac{\partial\lambda}{\partial p_k} &= \Phi_i^* \frac{\partial L_{ij}}{\partial p_k} \Phi_j = \Phi_k^* \Phi_k \quad , \qquad \frac{\partial\mu}{\partial p_k} = \Psi_k^* \Psi_k \\
\frac{\partial\lambda}{\partial q_k} &= i\Phi_j^* \frac{\partial f(q_j - q_l)}{\partial q_k} \phi_l = i \sum_l{}' f'(q_k - q_l)(\Phi_k^* \Phi_l - \Phi_l^* \Phi_k) \\
\cdots \quad \frac{\partial\mu}{\partial q_k} &= i \sum_l{}' f'(q_k - q_l)(\Psi_k^* \Psi_l - \Psi_l^* \Psi_k) \quad .
\end{aligned}
\tag{2.26}
$$

Hence

$$\{\lambda, \mu\} = i \sum_{k \neq l} f'(q_k - q_l) \{\Phi_k^* \Psi_k^* R_{kl} - \Phi_k \Psi_k R_{kl}^*\} \tag{2.27}$$

$$R_{kl} := \Phi_k \Psi_l - \Phi_l \Psi_k = -R_{lk} \quad .$$

In order to rewrite (2.27) in a form in which it is possible to explicitly use the crucial (functional) equation, (2.18), one notes that ($\lambda \neq \mu$)

$$\Psi_k \Phi_k = \frac{i}{\lambda - \mu} \sum_l f(q_k - q_l) R_{lk} \quad , \tag{2.28}$$

due to (2.23), i.e.

$$\lambda \Phi_k = p_k \Phi_k + i F_{kl} \Phi_l \quad , \quad \mu \Psi_k = p_k \Psi_k + i F_{kl} \Psi_l \tag{2.29}$$

implying

$$(\lambda - \mu)(\Phi_k \Psi_k) = i \sum_l (\Psi_k F_{kl} \Phi_l - \Phi_l F_{kl} \Psi_l)$$

$$= -i \sum_l f(q_k - q_l) R_{kl} \quad . \tag{2.30}$$

This allows one to write down an expression for the Poisson bracket $\{\lambda, \mu\}$ which after a moderately long chain of simple manipulations can be shown to be identically zero – without knowing the eigenvectors explicitly, i.e. only using the functional equation (2.18), and some symmetry properties of f and z,

$$f(-x) = -f(x) \quad , \quad z(-x) = z(x) \quad . \tag{2.31}$$

More explicitly, inserting (2.28) into (2.27), one obtains

$$\{\lambda, \mu\} = \frac{1}{\lambda - \mu} \sum_{\substack{l \neq k \\ j \neq k}} f'(q_k - q_l) \left\{ f(q_k - q_j) R_{jk} R_{kl}^* + f(q_k - q_j) R_{jk}^* R_{kl} \right\} \tag{2.32}$$

and (using $f'(x) = f'(-x)$, and interchanging j and l in the sum arising from the second term inside the curly bracket)

$$\{\lambda, \mu\} = -\frac{1}{\lambda - \mu} \sum{}'' \left(f' \underbrace{(q_k - q_l)}_{y} f \underbrace{(q_j - q_k)}_{x} - f'(q_j - q_k) f(q_k - q_l) \right) R_{jk} R_{kl}^* \quad . \tag{2.33}$$

Now using (2.18) one has

$$\{\lambda, \mu\} = \frac{1}{\mu - \lambda} \sum{}'' \left(z(q_j - q_k) - z(q_l - q_k) \right) f(q_j - q_l) R_{jk} R_{kl}^* \quad . \tag{2.34}$$

In order to go further, one notes that

$$\sum_l F_{jl}(\Phi_l \Psi_k - \Phi_k \Psi_l) = (F\Phi)_j \Psi_k - \Phi_k (F\Psi)_j$$

$$= -i \left\{ (\lambda \Phi_j - p_j \Phi_j) \Psi_k - \Phi_k (\mu \Psi_j - p_j \Psi_j) \right\} \tag{2.35}$$

implies

$$\sum_l f(q_j - q_l) R_{lk}^* = i\lambda \Psi_k^* \Phi_k^* - i\mu \Phi_k^* \Psi_j^* - i p_j R_{jk}^* \quad \text{(and c.c.)} \quad . \tag{2.36}$$

Hence

$$
\begin{aligned}
\{\lambda, \mu\} = \frac{1}{\mu - \lambda} \Bigg\{ &\sum_{j \neq k} z(q_j - q_k) R_{jk} \left[-i\Psi_k^* \Phi_j^* + i\mu \Phi_k^* \Psi_j^* + ip_j R_{jk}^* \right] \\
&+ \sum_{l \neq k} z(q_l - q_k) R_{kl}^* \left[-i\Psi_k \Phi_l + i\mu \Phi_k \Psi_l + ip_l R_{lk} \right] \Bigg\} \quad,
\end{aligned}
\tag{2.37}
$$

and (changing l into j in the second sum),

$$
\begin{aligned}
\{\lambda, \mu\} = -\frac{i\lambda}{\mu - \lambda} &\sum_{j \neq k} z(q_j - q_k) \left[R_{jk} \Psi_k^* \Phi_j^* + R_{kj}^* \Psi_k \Phi_j \right] \\
+ \frac{i\mu}{\mu - \lambda} &\sum_{j \neq k} z(q_j - q_k) \left[R_{jk} \Phi_k^* \Psi_j^* + R_{kj}^* \Phi_k \Psi_j \right] \quad.
\end{aligned}
\tag{2.38}
$$

As the quantities inside the two brackets in (2.38) are antisymmetric with respect to interchanging j and k,

$$
\begin{aligned}
\left[R_{jk} \Phi_k^* \Psi_j^* + R_{kj}^* \Phi_k \Psi_j \right] &= (\Phi_j \Psi_k - \Phi_k \Psi_j) \Phi_k^* \Psi_j^* + (\Phi_k^* \Psi_j^* - \Phi_j^* \Psi_k^*) \Phi_k \Psi_j \\
&= \Phi_j \Psi_k \Phi_k^* \Psi_j^* - \Phi_j^* \Psi_k^* \Phi_k \Psi_j \quad,
\end{aligned}
\tag{2.39}
$$

while $z(q_j - q_k)$ is symmetric, one finally has proven that

$$
\{\lambda, \mu\} = 0 \quad.
\tag{2.40}
$$

Notes and References

The proof of Poisson commutativity is taken from [5]. The main idea is analogous to that originally used in [6], [7] to prove the involutivity of the constants of motion for the Toda lattice (see **7, 8**, where an alternative proof is given). Yet another way of proving Poisson commutativity of integrals of motion is given in **11**. For an attempt to generalize the Calogero Moser systems to include different types of particles, see e.g. [2] or (for a rather complete discussion of the 3-body case) [8].

[1]   P.D. Lax; Comm. Pure Appl. Math. 21 (1968) 467.

[2]   F. Calogero; Lett. Nuov. Cim. 13 (1975) 411.

[3]   J. Moser; Adv. Math. 16 (1975) 197.

[4]   I.S. Gradshteyn, I.M. Ryzhik; *Tables of Integrals Series and Products*, Academic Press 1965.

[5]   A.M. Perelomov; *Integrable Systems of Classical Mechanics and Lie Algebras*, Birkhäuser 1990.

[6]   H. Flaschka; Phys. Rev. B9 (1974).

[7]   S. Manakov; Sov. Phys. JETP 40 (75) 269.

[8]   J. Hoppe, S. Theisen; Lett. Math. Phys. Vol. 22 #3 (1991) 229.

[9]   F. Calogero; Lett. Nuov. Cim. 16 (1976) 77.

[10]   M.A. Olshanetsky, A.M. Perelomov; Inv. Math. 37 (1976) 93.

[11]   I.M. Krichever; Funct. Anal. Appl. 14 (1980) 282.

3 Solution of a Quantum Mechanical N-Body Problem

Let us try to find the spectrum and Eigenfunctions of

$$H = -\frac{1}{2}\sum_{i=1}^{N}\frac{\partial^2}{\partial x_i^2} + \frac{1}{2N}\sum_{i>j}(x_i - x_j)^2 + g\sum_{i>j}\frac{1}{(x_i - x_j)^2} \quad . \tag{3.1}$$

First of all, (3.1) has to be supplemented by the class of functions it should act on. As the total momentum operator P commutes with (3.1) (due to the potential V depending only on the differences of the particle coordinates) the eigenfunctions ψ of (3.1) may be chosen to be also eigenfunctions of P; let us restrict ourselves to functions satisfying

$$P\psi = 0 \quad ; \quad P = -i\sum_{j=1}^{N}\frac{\partial}{\partial x_j} \quad . \tag{3.2}$$

Further, due to the singularity of V at the points $\mathbf{x} \in \mathbb{R}^N$ where two of the coordinates are equal, we require ψ to satisfy

$$\psi(x_1,\ldots,x_i,\ldots,x_j,\ldots,x_N) = 0 \quad \text{if } x_i = x_j \quad . \tag{3.3}$$

In fact, let us restrict ourselves to the domain $D = \{\mathbf{x} \in \mathbb{R}^N | x_1 < x_2 < \ldots < x_N\}$ on which ψ should be square integrable. Certainly, (3.3) will be satisfied, if ψ is proportional to

$$z := \prod_{i>j}(x_i - x_j) \tag{3.4}$$

– or some power of it.

Due to the term

$$r^2 := \frac{1}{N}\sum_{i>j}(x_i - x_j)^2 \tag{3.5}$$

in the potential – suggesting a factor $e^{\frac{-r^2}{2}}$ (possibly, times some polynomial in r) – one is lead to the Ansatz [1]

$$\psi(\mathbf{x}) = z^\epsilon \phi(r) Q(\mathbf{x}) \quad . \tag{3.6}$$

As a start, one may put $Q(x) = 1$ and/or $\phi(r) = e^{\frac{-r^2}{2}}$, but it is actually not much more work to leave ϕ and Q general when calculating $H\psi$:

$$\partial_i \psi = \epsilon z^{\epsilon-1}(\partial_i z)\phi Q + z^\epsilon(\partial_i r)\phi' Q + z^\epsilon \phi(\partial_i Q)$$

$$\partial_i^2 \psi = \epsilon(\epsilon-1)z^{\epsilon-2}(\partial_i z)^2 \phi Q + \epsilon z^{\epsilon-1}(\partial_i^2 z)\phi Q + 2\epsilon z^{\epsilon-1}Q(\partial_i z)(\partial_i r)\phi'$$
$$+ z^\epsilon \phi''(r)(\partial_i r)^2 Q + z^\epsilon \phi'(\partial_i^2 r)Q + 2z^\epsilon \phi'(\partial_i r)(\partial_i Q) \qquad (3.7)$$
$$+ \phi[(\partial_i^2 Q)z^\epsilon + 2\epsilon z^{\epsilon-1}(\partial_i z)(\partial_i Q)] \quad .$$

Obviously one needs to calculate the derivatives $\partial_i r$, $\partial_i z$ $\partial_i^2 r$, $\partial_i^2 z$, and sums of their mutual products:

From (3.5):

$$r(\partial_i r) = x_i - x \quad , \quad X = \frac{1}{N}\sum_j x_j \quad ; \qquad (3.8a)$$

applying once more ∂_i:

$$\sum_i \left((\partial_i r)^2 - r(\partial_i^2 r)\right) = N - 1 \quad ; \qquad (3.8b)$$

on the other hand:

$$\sum_i (\partial_i r)^2 = \frac{1}{r^2}\sum_i (x_i - X)^2 = 1 \quad , \qquad (3.8c)$$

as $\sum(x_i - X)^2 = \sum x_i^2 - NX^2$. Finally, $r^2 = \frac{1}{2}N\sum_{i,j}(x_i - x_j)^2 = \sum(x_i^2 - NX^2)$, which implies (using (3.8b))

$$\sum_i (\partial_i^2 r) = \frac{1}{r}(N - 2) \quad . \qquad (3.8d)$$

Differentiating the logarithm of (3.4) one finds

$$\partial_i z = z \sum_k{}' \frac{1}{x_i - x_k} \quad . \qquad (3.9a)$$

Hence

$$\sum_i (\partial_i z)(\partial_i r) = \frac{z}{r}\sum_{i\neq k}\frac{x_i - X}{x_i - x_k} = \frac{z}{r}\frac{N(N-1)}{2}$$

$$\left(\text{as} \quad \sum_{i\neq k}\frac{x_i - X}{x_i - x_k} = N(N-1) + \sum_{i\neq k}\frac{x_k - X}{x_i - x_k}\right) \quad , \qquad (3.9b)$$

and using

$$Y := \sum_{i\neq l\neq k\neq i}{}''' z_{ik}z_{il} \equiv 0 \quad ; \quad z_{ik} := \frac{1}{x_i - x_k} = -z_{ki} \quad , \qquad (3.9c)$$

$$\sum_i (\partial_i z)^2 = z^2 \sum_i \left(\sum_j{}' \frac{1}{x_i - x_j} \right) \left(\sum_k{}' \frac{1}{x_i - x_k} \right)$$

$$= z^2 \left[\sum_{i \neq l \neq k \neq i}{}''' z_{ij} z_{ik} + \sum_{i \neq k} \frac{1}{(x_i - x_k)^2} \right] \qquad (3.9d)$$

$$= z^2 \sum_{i \neq k} \frac{1}{(x_i - x_k)^2}$$

$$\sum_i (\partial_i^2 z) = z \sum_i \left(\sum_k{}' \frac{1}{x_i - x_k} \right)^2 - z \sum_{i \neq k} \frac{1}{(x_i - x_k)^2}$$

$$= z \left[\sum_i \left(\sum_j{}' \sum_k{}' \frac{1}{x_i - x_j} \frac{1}{x_i - x_k} - \sum_k{}' \frac{1}{(x_i - x_k)^2} \right) \right] \qquad (3.9e)$$

$$= z \sum{}''' z_{ij} z_{ik} = 0$$

(Eq. (3.9c) follows from

$$z_{ki} z_{il} + z_{il} z_{lk} + z_{lk} z_{ki} \equiv 0 \quad , \qquad (3.9f)$$

so that $Y = -2Y$ (by changing indices), hence $Y = 0$.) The N-particle Schrödinger-equation

$$H\psi = E\psi \qquad (3.10)$$

– with ψ as in (3.6) – therefore reads:

$$-\frac{1}{2} \Bigg\{ \epsilon(\epsilon - 1) \sum_{i \neq k} \frac{1}{(x_i - x_k)^2} \phi + 0 + N(N-1)\epsilon \frac{\phi'}{r} + \phi''$$

$$+ \frac{\phi'}{r}(N-2) \Bigg\} z^\epsilon Q - \frac{1}{2} \Bigg\{ 2\frac{\phi'}{r} \sum_i (x_i - X)\partial_i Q \Bigg\} z^\epsilon \qquad (3.11)$$

$$- \frac{1}{2}(DQ)z^\epsilon \phi + \left(\frac{1}{2}r^2 + g \sum_{i>j} \frac{1}{(x_i - x_j)^2} \right) \phi z^\epsilon Q$$

$$\overset{!}{=} E(\phi z^\epsilon Q) \quad ,$$

where

$$D := \sum_i \partial_i^2 + \epsilon \sum_{i \neq k} \frac{\partial_i - \partial_k}{x_i - x_k} \quad . \qquad (3.12)$$

Supposing

$$DQ = 0 \qquad (3.13)$$

$$\sum_i x_i \partial_i Q = k Q \quad , \qquad (3.14)$$

Eq. (3.11) reduces to an effective 1-dimensional 'Schrödinger–equation' for ϕ,

$$-\frac{1}{2}\phi'' - \frac{1}{2}\frac{\phi'}{r}\{N(N-1)\epsilon + (N-2) + 2k\} + \frac{1}{2}r^2\phi \overset{!}{=} E\phi \quad , \tag{3.15}$$

provided

$$\epsilon(\epsilon - 1) = g \quad , \quad \text{i.e.} \quad \epsilon = \frac{1}{2} + \sqrt{\frac{1}{4} + g} \quad . \tag{3.16}$$

Note that $\sum_i \partial_i Q = 0$ was implied by $P\psi = 0$, and that the subtle part of the reduction to (3.15) is the cancellation of the anharmonic part of the potential by terms appearing in the kinetic energy due to the factor z^ϵ.

Writing $\phi = e^{\frac{-r}{2}}P(r)$ yields

$$P'' + \left(\frac{\lambda}{r} - 2r\right)P' + (2E - \lambda - 1)P = 0 \tag{3.17}$$

$$\lambda = \lambda(k) = N(N-1)\epsilon + (N-2) + 2k \quad ,$$

and with $P(r) = L(r^2) = L(u)$,

$$uL'' + L'\left(\frac{\lambda+1}{2} - u\right) + \left(\frac{E}{2} - \frac{\lambda+1}{4}\right)L = 0 \quad . \tag{3.18}$$

If $\frac{E}{2} - \frac{\lambda+1}{2} = n \in \mathbb{N}_0$, i.e.

$$E = E_n = 2n + k + \frac{N-1}{2}(N\epsilon + 1) \qquad (n = 0, 1, 2, \ldots) \quad , \tag{3.19}$$

Eq. (3.18) has polynomial solutions $L_n^{\frac{\lambda-1}{2}}(u)$, called Laguerre polynomials [2]. So

$$\psi_{n,k}(\mathbf{x}) = z^\epsilon e^{\frac{-r^2}{2}} L_n^{\frac{\lambda-1}{2}}(r^2)Q_k(\mathbf{x}) \tag{3.20}$$

will be normalizable Eigenfunctions of (3.1), provided Q_k satisfies (3.13) (we have already used (3.14), which means that Q_k should be a homogeneous polynomial of degree k).

It remains to discuss (3.13), and to address the question, whether *all* solutions ψ have to be of the form (3.6), with Q satisfying (3.13) and (3.14).

For the discussion of (3.12) (as well as for any other differential equation $\tilde{D}\psi = 0$ with translational invariance, $[\tilde{D}, P] = 0$) it is convenient to introduce

$$x_i' := x_i - X \qquad \left(X = \frac{1}{N}\sum_{j=1}^{N}x_j\right) \tag{3.21}$$

$$\partial_i' := \partial_i - \frac{1}{N}\sum_{j=1}^{N}\partial_j \quad , \tag{3.22}$$

which satisfy

$$\sum_{i=1}^{N}x_i' = 0 \quad , \quad \sum_{i=1}^{N}\partial_i' = 0 \quad . \tag{3.23}$$

Note that for $N > 3$ it is *not* possible to directly eliminate the center of mass motion, as the number of pairs $x_i - x_j$ exceeds the number of independent coordinates; also note, that (3.22) - despite its suggestive notation - is a *definition* (it does not follow from any kind of chain rule); in particular, one has

$$\partial_i' x_j' = \delta_{ij} - \frac{1}{N} \tag{3.24}$$

instead of δ_{ij}. As $\partial_i' Q = \partial_i Q$ on translationally invariant Q, and $x_i' - x_j' = x_i - x_j$ anyway, (3.13) does not change its form when written in primed variables:

$$\left(\sum_i \partial_i'^2 + \epsilon \sum_{i \neq j} \frac{\partial_i' - \partial_j'}{x_i' - x_j'} \right) Q(\mathbf{x}') = 0 \quad . \tag{3.13'}$$

As one can show [1] that polynomial solutions of (3.13) – hence (3.13') – have to be completely symmetric functions of their arguments, one can explicitly eliminate the constraints on Q and the coordinates, by introducing the independent (unconstrained) variables

$$s_p := \sum_{i=1}^{N} (x_i')^p \quad , \quad p = 2, 3, \ldots, N \, .$$
$$\left(s_0 = N \, , \, s_1 = 0 \, , \, s_2 = r^2 \right) \quad . \tag{3.25}$$

$\sum_i x_i' = 0$ simply means that Q depends only on the s_p with $p > 1$. The resulting differential equation (written in terms of $s_{p>1}$ and $\partial/\partial s_p$, $p > 1$) can be found in [3], where it has been solved for $k < 6$ (note that there is only *one* solution for each $k = 3, 4, 5$, and none for $k = 2$):

$$Q_3 = s_3 = \sum (x_i')^3$$
$$Q_4 = (N + 1 + N(N-1)\epsilon) \, s_4 - \left(3 \left(1 - \frac{1}{N} \right) + (2N - 3)\epsilon \right) s_2^2 \tag{3.26}$$
$$Q_5 = (N + 5 + N(N-1)\epsilon) \, s_5 - 5 \left(2 \left(1 - \frac{1}{N} \right) + (N - 2)\epsilon \right) s_3 s_2 \quad .$$

$Q_3(Q_4 \ldots)$, written in terms of the x_i, would not be given by such simple expressions.

In general, there are as many independent solutions for fixed k as there are nonnegative integers n_p $(p > 2)$ satisfying

$$k = 3n_3 + 4n_4 + \ldots + N n_N \quad , \tag{3.27}$$

– e.g. 2 solutions for $k = 6, 7$ (3 for $k = 8, \ldots$). Denoting this number (of solutions) by $g(k, N)$, one has

$$\sum_{k=0}^{\infty} g(N, k) t^k = \frac{1}{(1 - t^3)(1 - t^4) \cdots (1 - t^N)} \quad . \tag{3.28}$$

The total number of states of Energy

$$E_m = m + E_0 \quad , \quad E_0 = \frac{N-1}{2}(N\epsilon + 1) \quad , \tag{3.29}$$

cp. (3.19), is therefore $= f(m, N)$, where

$$\sum_{k=0}^{\infty} f(k, N)t^k = \frac{1}{(1 - t^2)(1 - t^3)\cdots(1 - t^N)} \quad . \tag{3.30}$$

Notes and References

The highly influential original paper is [1].

[1]   F. Calogero; J. Math. Phys. Vol 12 #3 (1971) 419.
[2]   I.S. Gradshteyn, I.M. Ryzhik; *Tables of Integrals, Series and Products* Academic Press 1965.
[3]   A.M. Perelomov; Theor. Math. Phys. 6 (1971) 263.

4 Algebraic Approach to $x^2 + \alpha/x^2$ Interactions

Consider

$$H = \frac{1}{2}\left(-\partial_x^2 + x^2 + \frac{2\alpha}{x^2}\right) \quad , \quad H\psi = E\psi \tag{4.1}$$

$$\alpha \geq 0 \quad , \quad 0 < x < \infty \quad , \quad \psi(0) = 0 \quad .$$

Let us recall the situation for $\alpha = 0$:

$$H = a^+ a + \frac{1}{2} \quad , \quad a = \frac{1}{\sqrt{2}}(\partial_x + x) \quad , \quad a^+ = \frac{1}{\sqrt{2}}(-\partial_x + x) \tag{4.2}$$

$$[H, a] = -a \quad , \quad [H, a^+] = a^+ \quad , \quad [a, a^+] = \mathbb{1} \quad .$$

Without taking into account the boundary condition at 0, the eigenfunctions are

$$\phi_m(x) = \frac{(a^+)^m}{\sqrt{m!}}\phi_0(x) = \gamma_m H_m(x)e^{\frac{-x^2}{2}} = (-1)^m \phi_m(-x) \quad , \tag{4.3}$$

where the $H_m(x)$ are Hermite polynomials. Because of the boundary condition $\phi(0) = 0$ only odd $m \ (= 2n + 1)$ are allowed:

$$\psi_n = \phi_{2n+1} \quad ; \quad \psi_0 = \phi_1 = \gamma_1 x e^{\frac{-x^2}{2}} \quad . \tag{4.4}$$

Creation and annihilation operators A, A^+ that connect the eigenstates ϕ_n are

$$A^+ = (a^+)^2 \quad \text{and} \quad A = a^2 \quad ,$$

satisfying

$$[H, A] = -2A \quad , \quad [H, A^+] = 2A^+ \quad , \quad [A, A^+] = 4H \quad . \tag{4.5}$$

These are the commutation relations of $\text{sl}(2, \mathbb{R}) \hat{=} \text{so}(2, 1)$ (consult p.27, if necessary).

Using the normalisation $A_3 = \frac{1}{2}H$, $A_- = \frac{1}{2}A$, $A_+ = \frac{1}{2}A^+$ one has

$$[A_3, A_\pm] = \pm A_\pm \quad , \quad [A_+, A_-] = -2A_3 \quad , \tag{4.6}$$

with Casimir $C_2 = A_+ A_- - A_3^2 + A_3 \left(= \frac{-3}{4} \text{ in the 'fundamental' representation} \right.$

$$A_+ \leftrightarrow \begin{pmatrix} 0 & 0 \\ -1 & 0 \end{pmatrix}, \ A_- \leftrightarrow \begin{pmatrix} 0 & 1 \\ 0 & 0 \end{pmatrix}, \ A_3 \leftrightarrow \frac{1}{2}\begin{pmatrix} -1 & 0 \\ 0 & +1 \end{pmatrix} . \left. \right) \text{ The value of the}$$

Casimir in the representation used in (4.5) is

$$C_2 = \frac{1}{4}\left(A^+A - H^2 + 2H\right)$$

$$= \frac{1}{4}\left(a^{+2}a^2 - \left(a^+a + \frac{1}{2}\right)^2 + 2\left(a^+a + \frac{1}{2}\right)\right)$$

$$= \frac{1}{4}\left\{a^{+2}a^2 - \left(\underbrace{a^+aa^+a}_{=a^{+2}a^2+a^+a} + \frac{1}{4} + a^+a\right) + 2a^+a + 1\right\} \tag{4.7}$$

$$= \frac{3}{16}\mathbb{1} \quad .$$

The representation (4.5) is 'unitary' and irreducible, and

$$H = 2A_3 = \begin{pmatrix} \frac{3}{2} & & & 0 \\ & \frac{7}{2} & & \\ & & \frac{11}{2} & \\ 0 & & & \ddots \end{pmatrix} \quad , \qquad \begin{matrix} E_n = 2n + \dfrac{3}{2} \\ n = 0, 1, \ldots \end{matrix} \quad . \tag{4.8}$$

The situation that H appears as one of the generators of some Lie group (representation) is referred to as 'dynamical' – or 'spectrum generating' symmetry.

When trying to maintain this situation for $\alpha \neq 0$, i.e. looking for operators B, B^+ satisfying

$$[H, B] = -2B \quad , \quad [H, B^+] = 2B^+$$
$$[B, B^+] = 4H \quad ; \quad B = a^2 + \alpha f(x) \tag{4.9}$$

one finds that this is possible, and that $V(x) = \frac{c}{x^2}$ is in fact the only potential that can be added to (4.2) such that (4.9) can hold:

$$H = (a^+a + \frac{1}{2}) + \alpha V(x) = H_0 + \alpha V$$

$$[H, B] = [H_0, a^2] + \alpha\left([V, a^2] + [H_0, f(x)]\right)$$

$$= -2a^2 + \alpha\left(\frac{1}{2}[V, \partial_x^2 + 2x\partial_x + \ldots] - \frac{1}{2}[\partial_x^2, f]\right) \tag{4.9'}$$

$$\overset{!}{=} -2\left(a^2 + \alpha f(x)\right)$$

$$\Rightarrow f(x) = -V(x) \quad \text{and} \quad [V, x\partial_x] = +2V \quad \Rightarrow V \sim \frac{1}{x^2}$$

$$\text{So } B = a^2 - \frac{\alpha}{x^2} \quad .$$

In order to obtain the groundstate of (4.1), one solves

$$B\psi_0 = \left(\frac{1}{2}\left(\partial_x^2 + x^2 + 2x\partial_x + 1\right) - \frac{\alpha}{x^2}\right)\psi_0 \overset{!}{=} 0 \quad . \tag{4.10}$$

Making the Ansatz

$$\psi_0 = x^\epsilon e^{\frac{-x^2}{2}}$$

one finds $\frac{1}{2}\epsilon(\epsilon - 1) = \alpha$, i.e.

$$\epsilon = \frac{1}{2} + \sqrt{\frac{1}{4} + 2\alpha} \quad ,$$

$$H\psi_0 = \left(-B + x^2 + x\partial_x + \frac{1}{2}\right)\psi_0 = \left(\epsilon + \frac{1}{2}\right)\psi_0 = E_0\psi_0$$

(4.11)

$$\psi_n \sim (B^+)^n \psi_0 \quad , \quad E_n = 2n + \epsilon + \frac{1}{2} \quad n = 0, 1, 2, \ldots \quad .$$

(4.12)

Moreover,

$$C_2 = \frac{1}{4}\left(B^+ B - H^2 + 2H\right)$$

$$= \frac{1}{4}\left\{\left(a^{+2} - \frac{\alpha}{x^2}\right)\left(a^2 - \frac{\alpha}{x^2}\right) - \left(a^+ a + \frac{1}{2} + \frac{\alpha}{x^2}\right)^2 + 2\left(a^+ a + \frac{1}{2} + \frac{\alpha}{x^2}\right)\right\}$$

$$= \left(\frac{3}{16} - \frac{1}{4}\epsilon(\epsilon - 1)\right)\mathbb{1} \quad .$$

(4.13)

So $\psi = (\psi_0, \psi_1, \ldots)$ transforms according to an (irreducible) representation of $so(2,1)$ and, using [2] one knows that

$$B^+\psi_n = (-1)2\sqrt{(n+1)(n+\epsilon+\frac{1}{2})}\,\psi_{n+1}$$

$$B\psi_n = (-1)2\sqrt{n(n+\epsilon+\frac{1}{2})}\,\psi_{n-1} \quad ,$$

(4.14)

from which

$$\psi_n = (-1)^n \frac{1}{2^n}\sqrt{\frac{\Gamma(\epsilon+\frac{1}{2})}{n!\,\Gamma(n+\epsilon+\frac{1}{2})}}(B^+)^n\psi_0$$

(4.14')

easily follows. Writing

$$\psi_n(x) = \gamma_n x^\epsilon e^{\frac{-x^2}{2}} P_n(x^2)$$

one also has (with $x^2 = u$)

$$\left(u\partial_u^2 + \left(\epsilon + \frac{1}{2}\right)\partial_u - u\partial_u + n\right)P_n(u) = 0 \quad ,$$

(4.15)

i.e.

$$P_n(u) = L_n^{\epsilon - \frac{1}{2}}(u) \quad ,$$

the Laguerre polynomial [3] of degree n.

Let us now look at the N-body problem. Motivated by the remarkable fact that the spectrum of the fully interacting N-body system described in 3 is linear, one may look for some underlying algebraic structures, that would necessitate this linearity.

These were conjectured (and partially proved) in [1]: Suppose that there exist operators $B_2, B_3, \ldots, B_N$ obeying

$$[H, B_p] = -pB_p \quad . \tag{4.16}$$

$$\psi_{\mathbf{n}} = \left(B_2^+\right)^{n_2} \left(B_3^+\right)^{n_3} \cdots \left(B_N^+\right)^{n_N} \psi_0 \tag{4.17}$$

would then have energy

$$E_{\mathbf{n}} = \sum_{p=2}^{N} p\, n_p + E_0 \quad , \tag{4.18}$$

leading to the same spectrum and degeneracy that we have previously found (cp. 3.29/3.30)) – provided the wavefunctions (4.17) are independent (and complete).

The existence of B_2, B_3 and B_4, as well as a procedure to determine the B_p for $p > 4$ was shown in [1]: Introduce

$$b_i := \partial_i' + x_i' \quad (i = 1, 2, \ldots, N) \quad , \tag{4.19}$$

where x_i' and ∂_i' are defined as in 3 (3.21/3.22). Then

$$b_i^+ = -\partial_i' + x_i' \quad , \quad [b_i, b_j^+] = 2\left(\delta_{ij} - \frac{1}{N}\right)$$

$$[b_i, b_j] = 0 = [b_i^+, b_j^+] \tag{4.20}$$

$$\sum_{i=1}^{N} b_i = 0 = \sum_{i=1}^{N} b_i^+$$

and (3.1), which in x_i coordinates (and acting on translationally invariant functions) looks like

$$H = \frac{1}{2} \sum_i \partial_i'^2 + \frac{1}{2} \sum_i x_i'^2 + g \sum_{i>j} \frac{1}{(x_i' - x_j')^2} \tag{4.21}$$

$$= H_0 + g\, V \quad ,$$

can be written as

$$H = \frac{1}{2}\left(\sum_{i=1}^{N} b_i^+ b_i + (N-1)\right) + g \sum_{i>j} \frac{1}{(x_i' - x_j')^2} \quad . \tag{4.22}$$

In analogy with the 2-body case (cp.(3.9)/(3.5)) one can define

$$B_2 = \frac{1}{2} \sum_i b_i^{\,2} - gV \quad , \tag{4.23}$$

and finds

$$[H, B_2] = -2B_2 \quad , \quad [H, B_2^+] = 2B_2^+ \quad , \quad [B_2, B_2^+] = 4H \quad . \tag{4.24}$$

The determination of the B_p for $p > 2$, however, requires more work [1]. Again, one uses as a guideline the case $g = 0$, where there exist (besides $A_2 = \frac{1}{2}\sum b_i^2$) mutually commuting operators

$$A_p = \sum_{i=1}^{N} b_i^p \qquad (p = 2, \ldots, N) \tag{4.25}$$

satisfying

$$[H_0, A_p] = -pA_p \quad, \tag{4.26}$$

and the set of completely antisymmetric wavefunctions is given by

$$\phi_{\mathbf{n}} = \left(A_2^+\right)^{n_2} \left(A_3^+\right)^{n_3} \cdots \left(A_N^+\right)^{n_N} \phi_0 \tag{4.27}$$

with corresponding energies

$$E_{\mathbf{n}} = E_0 + 2n_2 + 3n_3 + \ldots + Nn_N \quad, \tag{4.28}$$

where

$$\phi_0 = \frac{1}{\sqrt{1!2!\cdots N!}} \frac{1}{2^{\frac{N-1}{4}}} \prod_{i>j} \left(b_i^+ - b_j^+\right)|0\rangle \tag{4.29}$$

$$b_i|0\rangle = 0 \quad \forall_i$$

is the state of lowest energy (within the class of totally antisymmetric wavefunctions).

For $g > 0$ one therefore makes the Ansatz

$$B_3 = A_3 + g\left(F_{ij}^{(3)}b_ib_j + F_i^{(3)}b_i + F^{(3)}\right)$$

$$B_4 = A_4 + g()\quad, \quad \ldots \quad, \tag{4.30}$$

and solves for the $F_{ij}^{(p)}$.

Some Notes on Lie Algebras

Formally, an *Algebra* $\mathcal{A} = (\mathcal{A}, \mathbf{K}, +, \circ)$ is a vectorspace with an additional multiplication $\circ$ (= a ring with additional multiplication by elements of the field $\mathbf{K}$). $\mathcal{A}$ is called 'commutative' if $x \circ y = y \circ x$, 'associative' if $x \circ (y \circ z) = (x \circ y) \circ z, \ldots$.

Examples: The set of all $n \times n$ matrices, with $\circ$ = the usual matrix multiplication, polynomials $P_n(x)$, $\ldots$

A *Lie Algebra* $G = (G, \mathbf{K}, +, \circ)$ is an algebra where $\circ$ satisfies:

(1) $\qquad\qquad x \circ y = -(y \circ x)$

(2) $\qquad\qquad x \circ (y \circ z) + y \circ (z \circ x) + z \circ (x \circ y) = 0$

Usually, $x \circ y$ is denoted by $[x, y]$, and 'in most cases' $[\ ,\ ]$ is realized as the commutator $x \cdot y - y \cdot x$ where $\cdot$ is the multiplication in some underlying *associative* algebra. In this case, conditions (2) (called 'Jacobi-identity') and (1) are automatically satisfied.

Lie algebras arise naturally in physics through their relations to Lie groups, which often appear as symmetry groups of physical systems, and (in the quantum theory) manifest themselves by the existence of operators (representations of elements of the corresponding Lie algebra) that either commute with the Hamiltonian, or contain H as one of their generators.

What follows, are some heuristic notes concerning the relation between Lie groups and -algebras, as well as a certain class of representations of so(2,1) that is realized in the physical systems discussed above.

First, consider ordinary rotations in N-dimensional space: $\mathbf{x} \rightarrow R\mathbf{x}$. They form a group,

$$
\begin{aligned}
SO(N) \;=\; & \{\text{all real } N \times N \text{ matrices } R \text{ of det} = +1 \text{ that leave invariant the} \\
& \quad \text{length of all vectors } \mathbf{x} \in \mathbb{R}^n \} \\
=\; & \{\text{all } R \text{ that } \dots \text{ the inner product } \mathbf{x}^t \cdot \mathbf{y}\} \\
=\; & \{\text{all } R \text{ satisfying } R^t R = 1, \text{ det } R = 1\}
\end{aligned}
$$

$-^t$ denotes transposition. The group elements R depend 'smoothly' on continuous real parameters ϵ_a ($a = 1, ..., N(N-1)/2$) corresponding to the angles of rotation in the $N(N-1)/2$ independent planes.

$$
\begin{aligned}
& R(\mathbf{0}) = 1 \\
& R(\epsilon) = 1 + \sum_a \epsilon_a T_a + O(\epsilon^2) && \text{(infinitesimal rotation)} \\
& R(\epsilon) = exp\left(\sum_a T_a\right) && \text{(finite rotation)} \\
& T_a^t = -T_a && \text{(from } R^t R = 1\text{)} \\
& \text{Tr } T_a = 0 && \text{(from det } R = 1\text{)}
\end{aligned}
$$

The generators of infinitesimal rotations are therefore real antisymmetric traceless matrices; the set of all those is not closed under multiplication, but closed under *commutation*. This is a special case of a much more general fact: Suppose one has a set $\{D(\epsilon)\}$ of elements (matrices, operators,...) depending 'smoothly' on some real parameters ϵ_a ($a = 1, \dots, d$), closed under multiplication:

$$
D(\epsilon') \cdot D(\epsilon'') = D\left(\epsilon(\epsilon', \epsilon'')\right)
$$

with $D(0) = 1$, $D^{-1}(\epsilon) = D(-\epsilon)$. Expanding both sides according to

$$
D(\epsilon) = 1 + \epsilon_a T_a + \epsilon_a \epsilon_b T_{ab} + 0(\epsilon^2)
$$

one finds

$$
\epsilon'_a \epsilon''_b T_a T_b = \epsilon'_a \epsilon''_b M^c_{ab} T_c + (\epsilon'_a \epsilon''_b + \epsilon''_a \epsilon'_b) T_a b
$$

where (using $D(0) = 1$)

$$
\epsilon_a(\epsilon', \epsilon'') = \epsilon'_a + \epsilon''_a + \epsilon'_b \epsilon''_c M^a_{bc} + 0(\epsilon^2) \quad .
$$

As T_{ab} is symmetric (by definition) one finds that

$$
T_a T_b - T_b T_a =: [T_a, T_b] = f_{ab}{}^c T_c \quad ,
$$
$$
f^c_{ab} = M^c_{ab} - M^c_{ba} = 2M^c_{ab} \quad .
$$

The linear space $L = \{\sum_a c_a T_a \mid c_a \in \mathbb{R}\}$ together with the Lie-bracket $[,]$ is called 'Lie Algebra' (corresponding to the 'Lie group' $\{D(\epsilon)\}$). For precise definitions see e.g. [4].

Actually, one can show that $D(\epsilon) = \exp(\sum_a \epsilon_a T_a)$, which means in particular, that the information about the Lie group multiplication – which corresponds to the knowledge of $\epsilon^a(\epsilon', \epsilon'')$, $a = 1, \ldots, d$ – is already contained in the 'structure constants' f^c_{ab}. Analogous to the discussion of SO(N), one can define

$$\text{SO}(N, M) = \{\text{all } (N + M) \times (N + M) \text{ matrices S that leave invariant the}$$
$$\text{inner product } (\mathbf{x}, \mathbf{y}) := \mathbf{x}^t \Lambda \mathbf{y} = \sum_{i=1}^{N} x_i y_i - \sum_{i=N+1}^{N+M} x_i y_i\}$$

Writing $S = e^T$, $\Lambda S^t \Lambda S = 1$ becomes

$$e^{\Lambda T^t \Lambda} e^T = 1 \ , \quad \text{i.e. } \Lambda T^t \Lambda = -T \ ,$$
$$\Lambda = \text{diag}(1 \ldots 1, -1 \ldots -1) \ .$$

With $T = \begin{pmatrix} A_{N \times N} & B_{N \times M} \\ C_{M \times N} & D_{N \times N} \end{pmatrix}$, this means $A^t = -A$, $D^t = -D$, $B^t = C$.

For the simplest nontrivial case, SO(2,1),

$$T = \begin{pmatrix} 0 & \alpha & \beta \\ -\alpha & 0 & \gamma \\ \beta & \gamma & 0 \end{pmatrix} = \alpha T_3 + \beta T_2 + \gamma T_1$$

$$T_3 = \begin{pmatrix} 0 & 1 & 0 \\ -1 & 0 & 0 \\ 0 & 0 & 0 \end{pmatrix} \ , \quad T_2 = \begin{pmatrix} 0 & 0 & 1 \\ 0 & 0 & 0 \\ 1 & 0 & 0 \end{pmatrix} \ , \quad T_1 = \begin{pmatrix} 0 & 0 & 0 \\ 0 & 0 & 1 \\ 0 & 1 & 0 \end{pmatrix} \ ,$$

with the commutation relations

$$[T_1, T_2] = -T_3 \ , \quad [T_2, T_3] = T_1 \ , \quad [T_3, T_1] = T_2 \ . \tag{$*$}$$

As expected, they differ from those for so(3) only by a sign. A particular class of 'unitary' (to be precise: unitarizable) representations of $(*)$ is

$$T_+ = \begin{pmatrix} 0 & 0 & 0 & \ldots \\ a_1 & 0 & 0 & \ldots \\ 0 & a_2 & 0 & \ldots \\ \vdots & \vdots & \ddots & \ddots \end{pmatrix} \ , \quad T_- = \begin{pmatrix} 0 & a_1 & 0 & \ldots \\ 0 & 0 & a_2 & \ldots \\ 0 & 0 & 0 & \ddots \\ \vdots & \vdots & \vdots & \ddots \end{pmatrix} \ , \tag{$**$}$$

$$T_3' = \text{diag}(b_1, b_2, \ldots) \ , \quad b_n = n + k - 1 \ , \quad a_n = \sqrt{n^2 + n(2k - 1)} \ ,$$

where one has defined $T_\pm = i(T_1 \pm iT_2)$, $T_3' = iT_3$. $(**)$ is called 'unitarizable' (or 'unitary') as the corresponding generators T_a $(a = 1, 2, 3)$ are antihermitean, which makes the representation of the group elements, $e^{\epsilon_a T_a}$, unitary. Note that $k = 3/4 \hat{=} \epsilon = 1 \hat{=} \alpha = 0$ (cp. (4.6),(4.8)). For the Casimir operator in the representation $(**)$ one finds

$$C_2 = T_3^2 - T_1^2 - T_2^2 = T_+ T_- - {T_3'}^2 + T_3'$$

$$= -k(k-1)\mathbb{1} = \left(\frac{3}{16} - \frac{1}{4}\epsilon(\epsilon - 1) \right).$$

$$\uparrow_{2k = \epsilon + \frac{1}{2}}$$

Notes and references

For more details, see the original work [1].

[1]   A.M. Perelomov; Theor. Math. Phys. 6 (1971) 263.
[2]   V. Bargmann; Annals of Math. Vol 48 #3 (1947) 568.
[3]   I.S. Gradsteyn, I.M. Ryshik; *Tables of Integrals, Series and Products*, Academic Press 1965.
[4]   S. Helgason; *Differential Geometry and Symmetric Spaces*, Academic Press 1962.

5 Some Hamiltonian Mechanics

Let us start with rewriting Hamilton's equations for the n-dimensional motion of a point particle in some external potential V,

$$\dot{q}^i = \frac{\partial H}{\partial p_i} \quad , \quad \dot{p}_i = -\frac{\partial H}{\partial q^i} \qquad (i = 1, 2, \ldots, n)$$

$$H = \frac{\mathbf{p}^2}{2m} + V(\mathbf{q}) \quad ,$$

(5.1)

in the form

$$\dot{x}^\alpha = J^{\alpha\beta} \frac{\partial H}{\partial x^\beta} \qquad (\alpha = 1, 2, \ldots, 2n)$$

(5.2)

$$\left(\dot{x} = J \cdot \nabla H \quad , \quad J = \begin{pmatrix} 0 & \mathbb{1}_{n \times n} \\ -\mathbb{1}_{n \times n} & 0 \end{pmatrix} \right)$$

or

$$\dot{x}^\alpha = \{H, x^\alpha\} \qquad (\alpha = 1, 2, \ldots, d) \quad ,$$

(5.3)

– where the Poisson-bracket $\{\ ,\ \}$ of two differentiable functions of x is defined as

$$\{F, G\}(x) = \sum_{j=1}^{n} \left(\frac{\partial F}{\partial q^j} \frac{\partial G}{\partial p_j} - \frac{\partial F}{\partial p_j} \frac{\partial G}{\partial q^j} \right)$$

(5.4)

$$= \sum_{\alpha,\beta=1}^{2n=d} \omega^{\alpha\beta} \partial_\alpha F \partial_\beta G \quad .$$

(5.5)

$\{\ ,\ \}$ should be thought of as an additional operation 'o' defined on the ring $(\mathcal{F}(M), +, \cdot)$ of (smooth) functions $F : M \to \mathbb{R}$. At the moment, $M = \mathbb{R}^{2n}$ (which, in particular, is a vector space), but one of the important generalizations will be to consider arbitrary *manifolds* M instead. As it is defined in (5.4), $\{\ ,\ \}$ satisfies

$$\{F, G\} = -\{G, H\}$$

(5.6a)

$$\{\{F, G\}, H\} + \{\{G, H\}, F\} + \{\{H, F\}, G\} = 0$$

(5.6b)

$$\{FG, H\} = F\{G, H\} + \{F, H\}G \quad .$$

(5.6c)

As the x^α in (5.3) are simply coordinates on M (which is conventionally called the *phase space*), one would like to know what kind of coordinate transformations

$$y^\alpha = f^\alpha(x_1, \ldots, x_d)$$

(5.7)

will leave invariant the (form of the) equations of motion, independent of the particular Hamiltonian.

Denoting by A the matrix formed out of

$$A^{\alpha}{}_{\beta} = \frac{\partial f^{\alpha}(x)}{\partial x^{\beta}} \qquad (5.8)$$

one finds

$$\dot{y}^{\alpha} = A^{\alpha}{}_{\beta}\dot{x}^{\beta} = A^{\alpha}{}_{\beta}\{H, x^{\beta}\} = A^{\alpha}{}_{\beta}J^{\beta\gamma}\partial_{\gamma}H$$
$$= A^{\alpha}{}_{\beta}(J)^{\beta\gamma}\frac{\partial y^{\epsilon}}{\partial x^{\gamma}}\frac{\partial K}{\partial y^{\epsilon}} \quad \left(K(y) := H\left(x\left(y\right)\right)\right) \quad ;$$

so

$$\dot{y}^{\alpha} = \{K, y^{\alpha}\}_{y} = J^{\alpha\epsilon}\frac{\partial K}{\partial y^{\epsilon}} \qquad (5.9)$$

if

$$AJA^{t} = J \quad . \qquad (5.10)$$

(Equivalently, one could have used (5.2) to derive (5.10)).

Invertible transformations (5.7) satisfying (5.10) are called 'canonical' (and $K = K(y)$ will be the Hamiltonian function in the new cordinates). Obviously, they form a group – $(AB)J(AB)^{t} = A(BJB^{t})A^{t} = J$. If one restricts to *linear* transformations, this group is called the 'symplectic' group, and is denoted by $SP(2n, \mathbb{R})$. It is a Lie group.

As an example, consider new variables $y = (Q^{1}, \ldots, Q^{n}, P_{1}, \ldots, P_{n})$ with

$$Q^{i} = Q^{i}(\mathbf{q}) = B^{i}{}_{j}q^{j} \quad , \quad P_{i} = P_{i}(\mathbf{p}) \quad . \qquad (5.11)$$

Then A will be of the form

$$A = \begin{pmatrix} B & 0 \\ 0 & C \end{pmatrix} \qquad (5.12)$$

which leads to the condition

$$C = (B^{t})^{-1} \qquad (5.13)$$

when substituted into (5.10). One can easily check that such a transformation 'preserves Poisson brackets' (– as any A satisfying (5.10) does):

$$\{Q^{i}, Q^{j}\} = 0 = \{P_{i}, P_{j}\}$$
$$\{Q^{i}, P_{j}\} = \sum_{k=1}^{n}\frac{\partial Q^{i}}{\partial q^{k}}\frac{\partial P_{j}}{\partial p_{k}} = B^{i}{}_{k}(C^{t})^{k}{}_{j} = \delta^{i}{}_{j} \quad . \qquad (5.14)$$

Canonical transformations can be very useful when trying to solve the equations of motion of a Hamiltonian system. Suppose e.g. that there exist n Poisson commuting conserved quantities $F_{i} : \mathbb{R}^{2n} \times \mathbb{R} \rightarrow \mathbb{R}, i = 1, 2, \ldots, n,$

$$\{F_{j}, F_{k}\} = 0 \qquad \frac{\partial F_{i}}{\partial t} + \{F_{i}, H\} = 0 \qquad (*)$$
$$(H = H(\mathbf{p}, \mathbf{q}, t))$$

which are independent on the set

$$M_a = \{(\mathbf{q}, \mathbf{p}, t) \in \mathbb{R}^{2n} \times \mathbb{R} \mid F_i(\mathbf{p}, \mathbf{q}, t) = a_i = \text{const.}, \ i = 1, \dots, n\}$$

(this is *one* possible definition of 'complete integrability'). Then [1] one may define new canonical variables $P_i = F_i(\mathbf{q}, \mathbf{p})$, $Q_i = \dots$ (in a way that the transformation becomes canonical), in which the equations of motion on M_a become trivial (choosing e.g. $F_2 = H$, one will have $Q_i = \text{const.}$ for $i \neq 2$ and $Q_2(t) = t + \text{const.}$). Such systems are said to be 'Liouville integrable', or 'integrable by quadratures'. A generalization to arbitrary $2n$ dimensional symplectic manifolds was given by Arnold [2].

Let us now look at how (5.5) can be generalized, while keeping the algebraic properties (5.6): Allowing ω^{ij} to be functions of x^α (instead of being mere constants), one easily sees that (5.6a) and (5.6c) are automatically satisfied (as long as $\omega^{ij} = -\omega^{ji}$), while the 'Jacobi-identitiy' (5.6b) reads

$$\{\{F, G\}\} + \text{cycl.} = \omega^{\gamma\delta} \partial_\gamma \left(\omega^{\alpha\beta}(x) \partial_\alpha F \partial_\beta G\right) \partial_\delta H + \text{cycl.}$$

$$= \omega^{\gamma\delta} \omega^{\alpha\beta} \Big[\partial^2_{\alpha\gamma} F \partial_\beta G \partial_\delta H + \partial_\alpha F \partial^2_{\beta\gamma} G \partial_\delta H$$

$$+ \partial^2_{\alpha\gamma} G \partial_\beta H \partial_\delta F + \partial_\alpha G \partial^2_{\beta\gamma} H \partial_\delta F \tag{5.15}$$

$$+ \partial^2_{\alpha\gamma} H \partial_\beta F \partial_\delta G + \partial_\alpha H \partial^2_{\beta\gamma} F \partial_\delta G \Big]$$

$$+ \left(\omega^{\gamma\delta} \partial_\gamma \omega^{\alpha\beta} + \omega^{\gamma\alpha} \partial_\gamma \omega^{\beta\delta} + \omega^{\gamma\beta} \partial_\gamma \omega^{\delta\alpha}\right) \partial_\alpha F \partial_\beta G \partial_\gamma H \quad .$$

Changing $\beta \to \delta$, $\delta \to \alpha$, $\alpha \to \gamma$, $\gamma \to \beta$ in the first term (and making analogous changes in the second and third term) one finds that the first and sixth term (and the second and third, fourth and fifth) cancel (using the antisymmetry of $\omega^{\cdot\cdot}$) – which of course was already assumed when claiming (5.6b) for the case of constant $\omega^{\cdot\cdot}$ – while the resulting condition on ω is

$$\omega^{\gamma\delta} \partial_\gamma \omega^{\alpha\beta} + \omega^{\gamma\alpha} \partial_\gamma \omega^{\beta\delta} + \omega^{\gamma\beta} \partial_\gamma \omega^{\delta\alpha} = 0 \quad . \tag{5.16}$$

If the matrix formed out of the ω^{ij} is invertible (which often is assumed from beginning by requiring (M, ω) to be a symplectic manifold),

$$\omega^{\alpha\beta} \omega_{\beta\gamma} = \delta^\alpha{}_\gamma \quad , \tag{5.17}$$

Eq.(5.16) may be restated as

$$\partial_\alpha \omega_{\beta\gamma} + \partial_\beta \omega_{\gamma\alpha} + \partial_\gamma \omega_{\alpha\beta} = 0 \quad , \tag{5.16'}$$

which – in the language of differential forms – simply means, that ω should be a *closed* 2-form.

A class of important examples, justifying e.g. that one should *not* restrict oneself to *even* dimensional phasespaces M from the very beginning, and displaying the generality gained by allowing arbitrary Poisson-structures, is

$$\omega^{\alpha\beta} = \sum_{\gamma} f^{\alpha\beta}{}_{\gamma} x^{\gamma} \qquad (5.18)$$

(i.e. assuming $\omega^{\cdot\cdot}$ to depend linearly on the coordinates of M). Inserting (5.18) into (5.16) one finds

$$f^{\alpha\beta}{}_{\gamma} f^{\gamma\delta}{}_{\epsilon} + f^{\beta\delta}{}_{\gamma} f^{\gamma\alpha}{}_{\epsilon} + f^{\delta\alpha}{}_{\gamma} f^{\gamma\beta}{}_{\epsilon} = 0 \quad, \qquad (5.19)$$

i.e. that the $f^{\alpha\beta}{}_{\gamma}$ should be the structure constants (in some basis) of some d-dimensional Lie algebra. Considering the simplest example of this type, $M = \mathbb{R}^3$ ($\hat{=}\mathrm{so}(3)^*$, the dual space of the Lie algebra $\mathrm{so}(3)$),

$$\{x^{\alpha}, x^{\beta}\} = \epsilon^{\alpha\beta\gamma} x^{\gamma} \qquad (5.20)$$

and, to be specific,

$$H = \frac{1}{2} \sum_{\alpha=1}^{3} a_{\alpha} (x^{\alpha})^2 \quad, \qquad (5.21)$$

one finds as the resulting equations of motion:

$$\begin{aligned}
\dot{x}_1 &= -(a_2 - a_3)\, x_2 x_3 \\
\dot{x}_2 &= -(a_3 - a_1)\, x_3 x_1 \quad, \\
\dot{x}_3 &= -(a_1 - a_2)\, x_1 x_2
\end{aligned} \qquad (5.22)$$

which are the Euler equations for a free top (x_i corresponding to the components of the angular momentum in the body fixed frame, and a_i being the inverse principal values of inertia).

Multiplying the first equation in (5.22) by x_1, the second by x_2, the third by x_3, and taking the sum of the three equations one recovers the time independence of the angular momentum, $\mathbf{x}^2$. Actually, this is not a consequence of the specific Hamiltonian (5.21), but a general feature following from (5.20), as

$$\{x^{\alpha}, \mathbf{x}^2\} = 0 \qquad (\alpha = 1, 2, 3) \qquad (5.23)$$

– which implies that *any* Hamiltonian (*any* function of $\mathbf{x}$) Poisson commutes with $\mathbf{x}^2$. Fixing $\mathbf{x}^2 = r^2 = \mathrm{const.}$, one could introduce spherical angles θ and φ according to

$$\begin{aligned}
x_1 &= r \sin\theta \cos\varphi \\
x_2 &= r \sin\theta \sin\varphi \quad, \\
x_3 &= r \cos\theta
\end{aligned} \qquad (5.24)$$

and consider, from the beginning, Hamiltonian mechanics on the surface of the sphere, S^2, for which the corresponding Poisson structure is given by

$$\begin{aligned}
\{f(\theta,\varphi), g(\theta,\varphi)\} &:= \frac{1}{\sin\theta} \left(\frac{\partial f}{\partial\theta} \frac{\partial g}{\partial\varphi} - \frac{\partial f}{\partial\varphi} \frac{\partial g}{\partial\theta} \right) \\
&= \partial_{\mu} f \partial_{\phi} g - \partial_{\phi} f \partial_{\mu} g
\end{aligned} \qquad (5.25)$$

$$\left(\mu := -\cos\theta \in (-1, +1) \right) \quad.$$

So, in a way, μ and φ are canonically conjugate variables, like p and $q \in \mathbb{R}$ – the important difference being only the topology (in particular, that μ and φ are not globally defined coordinates on the whole phase space).

Algebraically, the step from (5.20) to (5.25) corresponds to dividing out of the Poisson-Lie-algebra $\mathcal{P}$ (all polynomials in x_1, x_2, x_3) the ideal generated by $(\mathbf{x}^2 - r^2)$,

$$I = (\mathbf{x}^2 - r^2)\mathcal{P} \quad . \tag{5.26}$$

This point of view becomes important when trying to quantize such theories whose phase space is curved (– only for $\mathbb{R}^{2n}$ we know that we should simply replace q_i and p_j by operators Q_i and P_j satisfying $[Q_i, P_j] = i\hbar\delta_{ij}$). Instead of (5.24) we could use (5.20) for 'quantization', replacing (commutative) functions $f(x_1, x_2, x_3)$ by 'operators' $f(X_1, X_2, X_3)$ – elements of the enveloping algebra U of so(3),

$$[X_\alpha, X_\beta] = \epsilon_{\alpha\beta\gamma}X_\gamma \quad . \tag{5.27}$$

Just as in the commutative case, one could divide out of U an ideal, generated by $X_1^2 + X_2^2 + X_3^2 - r^2$. This corresponds to going to a representation of so(3) in which the quadratic Casimir $C_2 = X_1^2 + X_2^2 + X_3^2$ takes the value r^2. In this sense U_r (the enveloping algebra in the representation 'r'), and the dynamics that can be formulated in U_r, are quantizations of the space of smooth functions on S^2, and its Hamiltonian mechanics.

Notes and references

For an introduction to the modern formulation of Hamiltonian mechanics, see [2],[3]. For Hamiltonian mechanics on coadjoint orbids see e.g. [4]. The relation between functions on the sphere and $U_r(\text{so}(3))$ was considered in [5]. Someone interested in a discussion of 'what is integrability' should consult the recent book of the same title, edited by V.E. Zakharov [6].

[1]   J. Liouville; *Note sur les équations de la dynamique*,
       J. Math. Pures Appl. 20 (1855) 137.
[2]   V.I. Arnold; *Mathematical Methods of Classical Mechanics*, Springer 1978.
[3]   R. Abraham, J. Marsden; *Foundations of Mechanics*, Benjamin-Cummings 1978.
[4]   A.G. Reyman, M.A. Semenov-Tian-Shansky; *Group Theoretical Methods in the Theory of Integrable Systems*, Encyclopaedia of Mathematical Sciences, Vol. 16 Springer 1989.
[5]   J. Hoppe; MIT Ph.D Thesis 1982.
[6]   V.E. Zakharov (ed.); *What is Integrability*, Springer Series in Nonlinear Science, Springer 1991.

6 The Classical Non-Periodic Toda Lattice

Consider a system of N particles whose dynamics is governed by

$$H = \frac{1}{2}\sum_{i=1}^{N} p_i^2 + g^2 \sum_{i=1}^{N-1} e^{2(q_i - q_{i+1})} \tag{6.1}$$

(note that any interaction of the form $\sum_i g_i e^{c(q_i - q_{i+1})}$ with $g_i > 0$ leads to (6.1) by rescaling and shifting q_i and H).

The equations of motion read

$$\dot{q}_i = p_i \quad , \quad \dot{p}_1 = -2g^2 e^{2(q_1 - q_2)}$$

$$\dot{p}_j = 2g^2 \left(e^{2(q_{j-1} - q_j)} - e^{2(q_j - q_{j+1})} \right) \quad (1 < j < N) \tag{6.2}$$

$$\dot{p}_N = 2g^2 e^{2(q_{N-1} - q_N)}$$

or

$$\dot{p}_i = 2\left(a_{i-1}^2 - a_i^2\right) \quad , \quad \dot{a}_i = (p_i - p_{i+1})a_i \quad , \tag{6.3}$$

when defining

$$a_i = g e^{q_i - q_{i+1}} \quad , \quad a_0 \equiv 0 \equiv a_N \quad . \tag{6.4}$$

While it is rather hopeless trying to solve the coupled nonlinear equations (6.2/3) directly, one can easily convince oneself that they are equivalent to

$$\dot{L} = [L, M] \quad , \tag{6.5}$$

$$L = \begin{pmatrix} p_1 & a_1 & 0 & \cdots & & 0 \\ a_1 & p_2 & a_2 & \ddots & & \vdots \\ 0 & a_2 & p_3 & \ddots & & 0 \\ \vdots & \ddots & \ddots & \ddots & & a_{N-1} \\ 0 & \cdots & 0 & & a_{N-1} & p_N \end{pmatrix}, \quad M = \begin{pmatrix} 0 & a_1 & 0 & \cdots & & 0 \\ -a_1 & 0 & a_2 & \ddots & & \vdots \\ 0 & -a_2 & 0 & \ddots & & 0 \\ \vdots & \ddots & \ddots & \ddots & & a_{N-1} \\ 0 & \cdots & 0 & & -a_{N-1} & p_N \end{pmatrix} \tag{6.6}$$

As a consequence of (6.5), $\mathrm{Tr}\, L^k$ will be time-independent, which in its turn implies that the eigenvalues λ_i of $L(t)$ do not change in time. Moser [5] used this fact to obtain information on the time dependence of q_i and p_i in the following, indirect while ingenious, way: Let

$$f(\lambda) = \left((\lambda \mathbb{1} - L)^{-1}\right)_{NN} \quad . \tag{6.7}$$

Writing $(\ )_{NN}$ as $e_N^t(\ )e_N$, $e_N = (0,\ldots,0,1) = r_i\,(r_i e_N)$, where r_i is the ith (normalized) eigenvector of L,

$$Lr_i = \lambda_i r_i \qquad \begin{array}{l} r_i \cdot r_j = \delta_{ij} \\ i, j = 1, \ldots, N \end{array} , \tag{6.8}$$

(6.7) can be seen to be equivalent to

$$f(\lambda) = \sum_{i=1}^{N} \frac{r_i^2(t)}{\lambda - \lambda_i} , \tag{6.9}$$

where r_i is the Nth component of the ith eigenvector of L.

In order to obtain information on the time dependence of r_i consider

$$L(t) = R(t)\Lambda R^{-1}(t) , \tag{6.10}$$

i.e. the diagonalization of L (which is real symmetric) by a real orthogonal matrix $R(t)$, which can (and will) be chosen such that its ith column is the ith eigenvector of L, $r_i(t)$. So $r^t = (r_1, r_2, \ldots, r_N)$ coincides with the bottom row of R. Note that $RR^t = 1\!\!1$ implies

$$\sum_{i=1}^{N} r_i^2 = 1 , \tag{6.11}$$

which was of course already inherent in (6.7/8), as (6.7) goes like $1/\lambda$ for $\lambda \to \infty$. In any case, differentiating (6.10) with respect to t, rediscovering (6.5), one finds that

$$M = -\dot{R}R^{-1} , \tag{6.12}$$

which implies

$$\dot{r}_i = (\dot{r}_i)_N = -(MR)_{Ni} = -\sum_j M_{Nj} R_{ji} = a_{N-1}(r_i)_{N-1} . \tag{6.13}$$

In order to express the r.h.s. of (6.13) in terms of the r_j, note that the $(N-1)^{\text{th}}$ component of (6.8) yields

$$a_{N-1}(r_i)_{N-1} = (\lambda_i - p_N)r_i , \tag{6.14}$$

while the Nth component of (6.10) yields

$$p_N = \sum_{j=1}^{N} \lambda_j r_j^2 . \tag{6.15}$$

Alltogether, one therefore finds

$$\dot{x}_i = 2x_i \left(\lambda_i - \sum_j \lambda_j x_j \right) \qquad \begin{array}{l} x_i = r_i^2 \\ \sum x_i = 1 \end{array} . \tag{6.16}$$

Note that the constraint $\sum x_i = 1$ is indeed consistent with the (system of) differential equation(s) (6.16), as

$$\left(\sum \dot{x}_i\right) = 2x(t)\left(\left(\sum x_i\right) - 1\right) \ ,$$

$$x(t) = \sum_{j=1}^{N} \lambda_j x_j(t) \ . \tag{6.17}$$

In order to solve (6.16), let us make the Ansatz

$$x_i(t) = e^{2\lambda_i t} f_i(t) \ . \tag{6.18}$$

Solving the resulting differential equation for the $f_i(t)$ in terms of $x(t)$, one finds

$$x_i(t) = x_i(0) e^{2\lambda_i t} e^{-2\int_0^t x(t')dt'} \ . \tag{6.19}$$

Fortunately, one does not need to solve for $x(t)$, as the constraint $\sum x_i = 1$ implies

$$e^{2\int_0^t x(t')dt'} = \sum_{j=1}^{N} x_j(0) e^{2\lambda_j t} \ . \tag{6.20}$$

So

$$x_i(t) = \frac{x_i(0) e^{2\lambda_i t}}{\sum_{j=1}^{N} x_j(0) e^{2\lambda_j t}} \ . \tag{6.21}$$

The next step then is to relate the dynamical variables to the (unknown) functions $x_i(t)$. In order to do so, one notes that 'Kramers rule' for the matrix elements of the inverse of $L(\lambda) := (\lambda \mathbb{1} - L)$ implies

$$f(\lambda) = (L(\lambda))_{NN}^{-1} = \frac{\Delta_{N-1}}{\Delta_N} \ , \tag{6.22}$$

where Δ_k is defined to be the determinant of the upper left $k \times k$ block of $L(\lambda)$. On the other hand,

$$\Delta_i = (\lambda - p_i)\Delta_{i-1} - a_{i-1}^2 \Delta_{i-2} \ , \tag{6.23}$$

because of the triangular nature of $L(\lambda)$. Converting (6.23) into a recursion formula for $s_i = \frac{\Delta_i}{\Delta_{i-1}}$,

$$s_i = (\lambda - p_i) - \frac{a_{i-1}^2}{s_{i-1}} \ , \tag{6.24}$$

one finally obtains an expression of $f(\lambda)$ as a continued fraction involving only λ, a_i and p_i:

$$f(\lambda) = \frac{1}{s_N} = \cfrac{1}{(\lambda - p_N) - \cfrac{a_{N-1}^2}{(\lambda - p_{N-1}) - \cfrac{a_{N-1}^2}{\cdots \ - \cfrac{a_1^2}{\lambda - p_1}}}} \ . \tag{6.25}$$

Equating equal powers of λ in the numerator/denominator of (6.9) and (6.25) (written as rational functions of λ), one finds relations involving only the dynamical variables and the $x_i(t)$. For small N these can easily be used to obtain $q_i(t)$ and $p_i(t)$ as functions of t.

Let us demonstrate this for the simplest case, $N = 2$: writing out (6.9) and (6.25) yields

$$p_1 = \lambda_1 x_2 + \lambda_2 x_1$$
$$p_1 p_2 - a_1^2 = \lambda_1 \lambda_2 \tag{6.26}$$
$$p_1 + p_2 = \lambda_1 + \lambda_2 \quad ;$$

hence, restricting to $p_1 + p_2 = 0$ ($\lambda_2 = -\lambda_1 < 0$):

$$a_1^2 = \frac{4x_0(1 - x_0)\lambda_1^2}{(x_0 e^{2\lambda_1 t} + (1 - x_0)e^{-2\lambda_1 t})} \quad , \quad x_0 = x_1(0) \quad , \tag{6.27}$$

or

$$q(t) = \ln\left(\frac{1}{2}\left(e^{2\lambda_1 t} + be^{-2\lambda_1 t}\right)\right) + c$$
$$b = \frac{1 - x_0}{x_0} \quad , \quad c = -\ln(\lambda_1\sqrt{b}) \quad , \quad q = q_2(t) - q_1(t) \quad . \tag{6.28}$$

Note that for $p(0) = 0$: $b = 1$ and $c = q(0)$. One may easily check that (6.28) solves

$$\ddot{q}(t) = 4e^{-2q(t)} \quad , \tag{6.29}$$

which is the equation of motion for

$$H = \left(\frac{1}{2}\operatorname{Tr} L^2 = \right) \quad p^2 + e^{-2q} \quad . \tag{6.30}$$

The energy of (6.28) is λ_1^2. Of course, in this simple example, one could have solved the equation of motion directly. Also note that for $N = 2$ it would have been easy to directly determine $r_i(t)$, as a function of q and p, by finding the eigenvectors of L explicitly.

For $N = 3$, let us only write down the analogue of (6.26), i.e. the identities that follow when equating (6.9) and (6.25):

$$p_1 + p_2 = (\lambda_1 + \lambda_2)(1 - x_1 - x_2) - x_1\lambda_1 - x_2\lambda_2$$
$$a_1^2 = (\lambda_1 + \lambda_2)(x_1\lambda_2 + x_2\lambda_1) - \lambda_1\lambda_2(1 - x_1 - x_2) + p_1 p_2$$
$$a_1^2 + a_2^2 = \lambda_1^2 + \lambda_2^2 + \lambda_1\lambda_2 - (p_1 + p_2)^2 + p_1 p_2$$
$$a_1^2(p_1 + p_2) = -\lambda_1\lambda_2(\lambda_1 + \lambda_2) + p_1 a_2^2 + p_1 p_2(p_1 + p_2)$$

(having put already $p_1 + p_2 + p_3 = 0 = \lambda_1 + \lambda_2 + \lambda_3$).

Letting E_{ij} be the matrix which is everywhere 0 except at the position (i, j), where it has $+1$ as its entry, (6.6) can also be written as

$$L = \sum_{i=1}^{N} p_i E_{ii} + \sum_{1}^{N-1} a_i \left(E_i + E_{-i} \right)$$

$$M = \sum a_i (E_i - E_{-i}) \quad , \quad E_i = E_{i,i+1} = E_{-i}^{\text{tr}}$$

This form, or rather its variant (provided $P = \sum p_i = 0$):

$$L = \sum_{1}^{N-1} \left(b_i H_i + a_i \left(E_i + E_{-i} \right) \right)$$

$$H_i = E_{ii} - E_{i+1,i+1}$$

$$b_1 = \frac{1}{N} \left((N-1) \, p_1 - p_2 - \ldots - p_N \right)$$

$$b_2 = \frac{1}{N} \left((N-2) \, (p_1 + p_2) - 2p_3 - \ldots - 2p_N \right)$$

$$\vdots$$

$$b_{N-1} = \frac{1}{N} \left((p_1 + p_2 + \ldots + p_{N-1}) - (N-1) \, p_N \right)$$

are a possible starting point for generalizations of (6.1).

Notes and references

Much information about Toda lattices, including original work and physical motivations can be found in [1]. The involutivity of sufficiently many integrals of motion (which were found in [2]) was shown in [3], [4].

[1]   M. Toda; *Theory of Nonlinear Lattices* Springer 1981.
[2]   M. Henon; Phys. Rev. B9 (1974) 1921.
[3]   H. Flaschka; Phys. Rev. B9 (1974) 1924.
[4]   S. Manakov; Sov. Phys. JETP 40 (1975) 269.
[5]   J. Moser; Adv. in Math. 16 (1975) 197.

7 r-Matrices and Yang Baxter Equations

One way to prove involutivity of a set of conserved quantities $Q_k = \operatorname{Tr} L^k$, $k = 1, 2, \ldots$, for a Lax equation $\dot{L} = [L, M]$, with L and M lying in some Lie algebra $\mathcal{G}$, is to find an element r in the tensor product $\mathcal{G} \otimes \mathcal{G}$ satisfying

$$\{L \overset{\otimes}{,} L\} = [r, L \otimes \mathbb{1}] - [r^\pi, \mathbb{1} \otimes L] \quad . \tag{7.1}$$

For simplicity, we will assume that $\mathcal{G}$ is finite dimensional and is realized as a Lie algebra of matrices acting on a finite dimensional (real or complex) vectorspace V with fixed orthonormal basis $\{e_i\}$, $i = 1, \ldots, N$. So L and M will simply be (real or complex) $N \times N$ matrices.

Before explaining the notation used in (7.1) and showing that it implies

$$\{\operatorname{Tr} L^k, \operatorname{Tr} L^l\} = 0 \quad \forall k, l \quad , \tag{7.2}$$

recall some rules for tensor products of associative algebras, like

$$(A \otimes B)(C \otimes D) = (AC) \otimes (BD) \quad ,$$
$$(A \otimes B) + (A \otimes C) = A \otimes (B + C) \quad ,$$
$$\widehat{\operatorname{Tr}}(A \otimes B) = (\operatorname{Tr} A) \cdot (\operatorname{Tr} B) \quad ,$$
$$(A \otimes B)_{ij,kl} = A_{ik} B_{jl} \quad .$$

The last line refers to an orthonormal basis e_i in V, $(e_i, e_j) = \delta_{ij}$, defining a basis E_{ij} in the space $\operatorname{End}(V)$ of linear maps from V to V via $E_{ij} e_k = \delta_{jk} e_i$, and writing $T \in \mathcal{G} \otimes \mathcal{G} \subset \operatorname{End}(V) \otimes \operatorname{End}(V) =: \mathcal{T}$ as

$$T = \sum T_{ij,kl} E_{ik} \otimes E_{jl} \quad , \tag{7.4}$$

i.e. letting $T_{ij,kl} = (T)_{ij,kl}$ denote the coefficient of $E_{ik} \otimes E_{jl}$; the trace $\widehat{\operatorname{Tr}}$ of $T \in \mathcal{T}$ is defined as $\widehat{\operatorname{Tr}} T = \sum_{m,n} (e_m \otimes e_n, T(e_m \otimes e_n)) = \sum_{ij} T_{ij,ij}$. The l.h.s. of (7.1) is understood as an element of $\mathcal{G} \otimes \mathcal{G} \subset \mathcal{T}$ whose ij, kl component (in the expansion (7.4)) is

$$\{L \overset{\otimes}{,} L\}_{ij,kl} := \{L_{ik}, L_{jl}\} \quad . \tag{7.5}$$

On the r.h.s., $r^\pi := \pi r \pi$, where the permutation operator $\pi \in \mathcal{T}$ is defined as $\pi(\mathbf{x} \otimes \mathbf{y}) := \mathbf{y} \otimes \mathbf{x}$. So

$$\pi_{ij,kl} = \delta_{il} \delta_{jk} \quad , \quad \pi^2 = \mathbb{1} \otimes \mathbb{1} \quad , \tag{7.6}$$

and

$$[r^\pi, \mathbb{1} \otimes L]_{ij,kl} = [r, L \otimes \mathbb{1}]_{ji,lk} \quad . \tag{7.7}$$

Note that both sides of (7.1) change sign under conjugation with π.

Let us now prove (7.2): Using the 'derivation-property' of the Poisson-bracket (and keeping track of the order),

$$\{f \cdot g, h\} = f \cdot \{g, h\} + \{f, h\} \cdot g$$

one obtains

$$\{L^k \overset{\otimes}{,} L^l\} = \sum_{i=0}^{k-1} \sum_{j=0}^{l-1} L^i \otimes L^j \left([r, L \otimes \mathbb{1}] - [r^\pi, \mathbb{1} \otimes L]\right) L^{k-i-1} \otimes L^{l-j-1} \qquad (7.8)$$

from (7.1). Taking the trace, using $\hat{\mathrm{Tr}}(T_1 T_2) = \hat{\mathrm{Tr}}(T_2 T_1)$ on the r.h.s, and $\mathrm{Tr}(A \otimes B) = (\mathrm{Tr}\, A)(\mathrm{Tr}\, B)$ on the l.h.s, one gets (7.2).

To actually find an r that satisfies (7.1) (for given L) is of course a far more difficult task (despite the fact, that one can also prove that (7.2) implies (7.1); the 'formula' for r, as given in [11], e.g., involves knowing the eigenvalues and -vectors of L explicitly, which is practically impossible). Writing out $[r, L \otimes \mathbb{1}]$ in components,

$$[r, L \otimes \mathbb{1}]_{ij,kl} = \sum_{\alpha,\beta} \left(r_{ij,\alpha\beta} (L \otimes \mathbb{1})_{\alpha\beta,kl} - (L \otimes \mathbb{1})_{ij,\alpha\beta}\, r_{\alpha\beta,kl}\right)$$
$$= \sum_{\alpha} \left(L_{\alpha k} r_{ij,\alpha l} - L_{i\alpha} r_{\alpha j,kl}\right) \quad , \qquad (7.9)$$

one finds that (7.1) is equivalent to

$$\{L_{ik}, L_{jl}\} = L_{\alpha k} r_{ij,\alpha l} - L_{i\alpha} r_{\alpha j,kl} - L_{\alpha l} r_{ji,\alpha k} + L_{j\alpha} r_{\alpha i,lk} \quad , \qquad (7.10)$$

where a sum over α (on the r.h.s.) is implied. Supposing L to be of the form

$$L_{\beta\gamma} = \delta_{\beta\gamma} p_\gamma + X_{\beta\gamma}(\mathbf{q}) \quad , \qquad (7.11)$$

i.e. assuming that the momenta enter (only) via the diagonal part of L (which is reasonable for translationally invariant systems, as the conservation of total momentum then comes out naturally as $\mathrm{Tr}\, L = \sum p_\gamma$) (7.10) reads:

$$\frac{\partial X_{ik}}{\partial q_j} \delta_{jl} - \delta_{ik} \frac{\partial X_{jl}}{\partial q_i} = r_{ij,kl}(p_k - p_i) - r_{ji,lk}(p_l - p_j)$$
$$+ X_{\alpha k} r_{ij,\alpha l} - X_{i\alpha} r_{\alpha j,kl}$$
$$- X_{\alpha l} r_{ji,\alpha k} + X_{j\alpha} r_{\alpha i,lk} \quad . \qquad (7.12)$$

In general, r will depend on the phase space variables q and p. The simplest possibility, however, would be a constant r-matrix, - or to have r at least p-independent. In this case the explicitly p-dependent terms in (7.12) have to cancel, i.e.

$$r_{ij,kl}(p_k - p_i) - r_{ji,lk}(p_l - p_j) = 0 \quad \forall_{ijkl} \quad . \qquad (7.13)$$

While the simplest solution of (7.13) would be to have $r_{ij,kl}$ proportional to δ_{ik}, other ones are

$$r_{ij,kl} = \delta_{ij}\delta_{kl}r_{ik}^+ \quad \left(r^\pi = \quad r\right)$$

$$r_{ij,kl} = \delta_{il}\delta_{jk}r_{ij}^- \quad \left(r^\pi = -r\right) \quad . \tag{7.14}$$

For the non–periodic Toda lattice, (cp. (6.6)),

$$L_{ij} = p_i\delta_{ij} + a_i\delta_{i+1,j} + a_{i-1}\delta_{i-1,j} \quad , \tag{7.15}$$

one finds (see below) that the Ansatz (7.14^-), because of similar index structures of the two sides of (7.12) for this case, leads to the simple equations

$$r_{ij}^- - r_{i,j+1}^- = \delta_{i,j+1} + \delta_{ij} \quad \begin{array}{l} i = 1,\ldots,N, \\ j = 1,\ldots,N-1 \end{array}, \tag{7.16}$$

with $r_{ij}^- = -r_{ji}^-$ (due to $r^\pi = -r$), which up to an overall constant, immediately implies $r_{ij}^- = +1$ for $i > j$, and $r_{ij}^- = -1$ for $i < j$. So

$$r_{ij}^- = \theta(i-j) - \theta(j-i) \quad . \tag{7.17}$$

Actually, one can prove that (apart from minor modifications) (7.15) is the *only* L of the form (7.11) (X purely off-diagonal) that satisfies (7.12) with r given by (7.14^-): Inserting (7.14^-) into (7.12), one gets

$$\delta_{il}X_{jk}(r_{ij} + r_{kl}) + \delta_{jk}X_{il}(r_{ij} + r_{kl}) \overset{!}{=} \delta_{il}\partial_j X_{ik} - \delta_{ik}\partial_i X_{il} \quad . \tag{*}$$

Setting $j = l$ yields a first order differential equation for X_{ik} whose general solution is

$$X_{ik} = \gamma_{ik}e^{(q_i-q_k)}r_{ki}^- \quad , \quad \gamma_{ik} = \text{const} \, .$$

Inserting this into (*) and setting $i = l$ implies

$$\gamma_{jk}(r_{ij} + r_{ki}) = 0 \quad \left(j \neq i \neq k\right) \quad ,$$

from which follows that

$$\gamma_{jk} = 0 \quad \text{for} \quad |j - k| > 1$$
$$r_{ij}^- = \theta(i-j) - \theta(j-i) \quad .$$

So $X_{ik} = a_i\delta_{i+1,k} + a_{i-1}\delta_{i-1,k}$. One should note that by analogous considerations one finds that (7.14^+) is incompatible with (7.12).

Finally, let us write out explicitly the result of inserting (7.15) and (7.14^-) into (7.12), while also using $\partial_i a_j = (\delta_{ij} - \delta_{i,j+1})a_j$:

$$(r_{ij} + r_{kl})\{\delta_{il}\left(a_j\delta_{j+1,k} + a_{j-1}\delta_{j-1,k}\right) + \delta_{jk}\left(a_i\delta_{i+1,l} + a_{i-1}\delta_{i-1,l}\right)\}$$
$$\overset{!}{=} \delta_{jl}\{(\delta_{ij} - \delta_{j,i+1})a_i\delta_{i+1,k} + (\delta_{j,i-1} - \delta_{ij})a_{i-1}\delta_{i-1,k}\}$$
$$- \delta_{ik}\{(\delta_{ij} - \delta_{i,j+1})a_j\delta_{j+1,l} + (\delta_{i,j-1} - \delta_{ij})a_{j-1}\delta_{j-1,l}\} \quad .$$

The terms proportional to δ_{il} are

$$(r_{ij} + r_{kl})\,(a_j\delta_{j+1,k} + a_{j-1}\delta_{j-1,k})$$

$$- \{\delta_{ij}a_j\delta_{i+1,k} - \delta_{i,j-1}a_{j-1}\delta_{ik} - \delta_{ij}a_{j-1}\delta_{i-1,k} + a_j\delta_{ik}\delta_{i,j+1}\}\ \left(\overset{!}{=} 0\right)\ ,$$

so that comparing the coefficients of a_j yields (7.16).

For the *periodic Toda lattice*

$$H = \frac{1}{2}\sum_{i=1}^{N} p_i^2 + \sum_{i=1}^{N} e^{2(q_i - q_{i+1})} \quad , \tag{7.18}$$

L, as given in (7.15) (but now with $N+1 \overset{\wedge}{=} 1$, $N \overset{\wedge}{=} 0$), will still satisfy $\dot{L} = [L, M]$, with

$$M = a_i\delta_{i+1,j} - a_{i-1}\delta_{i-1,j} \quad , \tag{7.19}$$

and the Ansatz (7.14^-) will again lead to (7.16) – but now including $j = N$! The additional equation

$$r_{1N}^- = +1 = -r_{N1} \quad , \tag{7.20}$$

however, turns out to be inconsistent with the other ones, which implied

$$r_{1N}^- = r_{1,N-1}^- = \cdots = r_{1,2} = +1 \quad . \tag{7.21}$$

So there does *not* exist an r-matrix of the form (7.14^-) if L is given by (7.15).

Instead of looking for an r-matrix of a different, e.g. phase-space dependent form, one notes that one can introduce a 'spectral' parameter λ in L and M,

$$L(\lambda) = \begin{pmatrix} p_1 & \lambda a_1 & 0 & \cdots & & 0 & \frac{a_N}{\lambda} \\ \frac{a_2}{\lambda} & p_2 & \lambda a_2 & \ddots & & \cdots & 0 \\ 0 & \ddots & \ddots & \ddots & & & \vdots \\ \vdots & & \ddots & & \lambda a_{N-2} & & 0 \\ 0 & \cdots & 0 & \frac{a_{N-2}}{\lambda} & \ddots & & \lambda a_{N-1} \\ \lambda a_N & 0 & \cdots & 0 & & \frac{a_{N-1}}{\lambda} & p_N \end{pmatrix} \quad , \tag{7.22\,a}$$

$$M(\lambda) = \begin{pmatrix} 0 & \lambda a_1 & 0 & \cdots & & 0 & -\frac{a_N}{\lambda} \\ -\frac{a_2}{\lambda} & 0 & \lambda a_2 & \ddots & & \cdots & 0 \\ 0 & \ddots & \ddots & \ddots & & & \vdots \\ \vdots & & \ddots & & \lambda a_{N-2} & & 0 \\ 0 & \cdots & 0 & -\frac{a_{N-2}}{\lambda} & \ddots & & \lambda a_{N-1} \\ \lambda a_N & 0 & \cdots & 0 & & -\frac{a_{N-1}}{\lambda} & 0 \end{pmatrix} \quad , \tag{7.22\,b}$$

$$\dot{L}(\lambda) = [L(\lambda), M(\lambda)] \quad . \tag{7.23}$$

such that (7.23) is still equivalent to the equations of motion corresponding to (7.18) (in fact, one can easily show that $\text{Tr}\, L(\lambda)^k$ does not depend on λ for *any* $k < N$).

The integrability of H would therefore follow if one found an $r = r(\lambda, \mu)$ satisfying

$$\{L(\lambda) \overset{\otimes}{,} L(\mu)\} = [r(\lambda, \mu), L(\lambda) \otimes \mathbb{1}] - [r(\lambda, \mu)^{\pi}, \mathbb{1} \otimes L(\mu)] \quad . \tag{7.24}$$

Using (7.22), i.e.

$$L_{ij}(\lambda) = p_i \delta_{ij} + \lambda a_i \delta_{i+1,j} + \frac{1}{\lambda} a_{i-1} \delta_{i-1,j} \quad , \tag{7.25}$$

and (7.14$^-$), one finds as the analogue of (7.16) the conditions

$$x r_{ij}^-(x) - r_{ij+1}^-(x) = x \delta_{ij} + \delta_{ij+1} \quad . \tag{7.26}$$

$$r_{ij}^-(x) = -r_{ij}^-\left(\frac{1}{x}\right) \quad ; \quad x := \frac{\lambda}{\mu}$$

For $x \neq 1$, these equations do have a solution: Fixing e.g. $i = 1$ (and dropping the minus sign in r_{ij}^-) one obtains

$$x r_{11} - r_{12} = x \quad (j = 1) \quad ,$$
$$r_{1N} = x r_{1N-1} = \ldots = x^{N-2} r_{12} \quad (\text{from } j = 2, \ldots, N-1)$$
$$x r_{1N} - r_{11}(x) = 1 \quad ,$$

so

$$r_{11}(x) = \frac{x^N + 1}{x^N - 1}$$
$$r_{12}(x) = \frac{2x}{x^N - 1} \quad , \quad r_{13}(x) = \frac{2x^2}{x^N - 1} \quad , \ldots, \quad r_{1N}(x) = \frac{2x^{N-1}}{x^N - 1} \quad . \tag{7.28}$$

Analogous considerations for $i \neq 1$, and/or symmetry arguments, then yield the final answer,

$$r_{ij}(x) = \frac{1}{x^N - 1} \left\{ \delta_{ij} \left(x^N + 1 \right) + 2\Theta(j - i) x^{j-i} + 2\Theta(i - j) x^{N-i+j} \right\} \quad . \tag{7.29}$$

To be sure, one can check that it satisfies (7.26).

Yang-Baxter Equations

Due to the Poisson-bracket on the l.h.s. of (7.1), respectively (7.24), any solution r will have rather special algebraic properties. To explore these, e.g. for the case $r^\pi = -r$ and r independent of phase space, introduce the notation

$$L_1 = L(\lambda_1) \otimes \mathbb{1} \otimes \mathbb{1} \quad , \quad L_2 = \mathbb{1} \otimes L(\lambda_2) \otimes \mathbb{1} \quad , \dots ,$$
$$r^{1,2} = r(\lambda_1/\lambda_2) \otimes \mathbb{1} \quad , \quad r^{2,3} = \mathbb{1} \otimes r(\lambda_2/\lambda_3) \quad , \dots ,$$

and consider the cyclic sum of

$$\{\{L_1, L_2\}, L_3\} = \left\{ [r^{1,2}, L_1 + L_2], L_3 \right\}$$
$$= [r^{1,2}, [r^{1,3}, L_1 + L_3]] + [r^{1,2}, [r^{2,3}, L_2 + L_3]] \quad . \tag{7.30}$$

What one then finds is that

$$\left[[r^{1,2}, r^{1,3}] + [r^{1,2}, r^{2,3}] + [r^{1,3}, r^{2,3}], L_1 + L_2 + L_3 \right]$$

must be zero. (Start with this expression; use the Jacobi identity for commutators, together with $r^{a\,b} = -r^{b\,a}$, and $[r^{a\,b}, L_c] = 0$ if $c \neq a, b$). A sufficient condition for

$$\{\{L_1, L_2\}, L_3\} + \{\{L_2, L_3\}, L_1\} + \{\{L_3, L_1\}, L_2\}$$

to be zero is, therefore,

$$[r^{1,2}(x), r^{1,3}(xy)] + [r^{1,2}(x), r^{2,3}(y)] + [r^{1,3}(xy), r^{2,3}(y)] = 0 \tag{7.31}$$

which is called the 'Classical Yang Baxter Equation' (CYBE). (If L does not depend on some spectral parameter, $x = \lambda_1/\lambda_2$ and $y = \lambda_2/\lambda_3$ are simply put equal to 1). Equation (7.31) may be considered as the $O(\hbar^2)$ content of the 'Quantum Yang-Baxter Equation' (QYBE),

$$R^{1,2}(x) R^{1,3}(xy) R^{2,3}(y) = R^{2,3}(y) R^{1,3}(xy) R^{1,2}(x) \quad , \tag{7.32}$$

which means that (7.31) is obtained from (7.32) by writing

$$R(x, \hbar) = \kappa(x, \hbar)(\mathbb{1} + \hbar r(x) + \dots) \tag{7.33}$$

and ignoring terms of order $\hbar^3$ (and higher). Formally, $R(x, \hbar)$ lies in the tensor product $U(\mathcal{G}) \times U(\mathcal{G})$, where $U(\mathcal{G})$ is the enveloping algebra of $\mathcal{G}$ (which one may think of as the associative algebra of polynomials in elements of $\mathcal{G}$, with two polynomials identified if they can be obtained from each other upon using the commutation relations of $\mathcal{G}$).

The QYBE, (7.32), plays a rather fundamental role, and has appeared in about as many contexts as it has led to new directions of research. One of the simplest *physical* ways to describe the appearance of such an equation can be scetched as follows (for more precise statements, see [9]): Consider some simple $1 + 1$ dimensional relativistic theory of N species of point particles. A *free* particle of type j would be characterized by giving its relativistic 2-momentum

$(p^0, p^1) = (m \cosh \Theta, m \sinh \Theta)$, where m is its mass and Θ its rapidity; by definition such a state would be an eigenstate of the 2-momentum operator P^μ. The scattering of two such particles would be described by a probability amplitude

$$S^{j_1 j_2}_{i_1 i_2}(\Theta_{12}) \quad , \tag{7.34}$$

where Θ_{12} is the relative rapidity, i_1 and i_2 are the types of incoming particles, j_1 and j_2 those of the outgoing ones. One could draw a picture

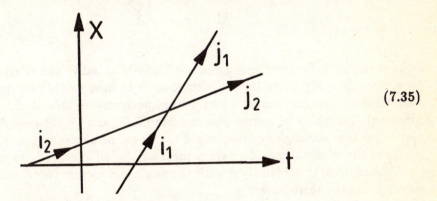

$$\tag{7.35}$$

to visualize the 'process' to which (7.34) is attached.

'Simple' theories now have the property that the S–matrix for multiple particle scattering 'factorizes', which means that the n-particle amplitudes

$$S^{j_1 \cdots j_n}_{i1 \cdots in}(\Theta_1, \ldots, \Theta_n)$$

can be written as products of two-particle scattering amplitudes, e.g.

$$S^{j_1 j_2 j_3}_{i_1 i_2 i_3}(\Theta_1, \Theta_2, \Theta_3) = \sum_{k_1 k_2 k_3} S^{k_1 k_2}_{i_1 i_2}(\Theta_{12}) S^{j_1 k_3}_{k_1 i_3}(\underbrace{\Theta_{13}}_{=\Theta_{12}+\Theta_{23}}) S^{j_2 j_3}_{k_2 k_3}(\Theta_{23}) \quad . \tag{7.36}$$

Visualizing (7.36) as

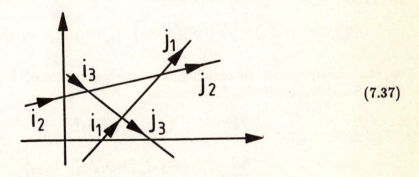

$$\tag{7.37}$$

suggests to say that only two particles scatter 'at a time', and that the 3-particle scattering which in principle could involve complicated 'processes' inside

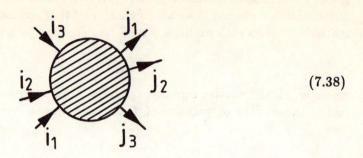

$$(7.38)$$

looks as in (7.37). However, using more or less only causality and relativistic invariance, one can easily prove that it is impossible to have (within the framework of a relativistic quantum theory) a well defined position operator. Although (strictly speaking) this makes of course pictures like (7.35) and (7.37) very dubious, one may argue that although position might not be defined, momentum *is*, so that the angles of the straight lines do have a meaning. So all one needs to do to render (7.37) consistent, is to identify it with the analogous picture (obtained by parallel transporting the three lines),

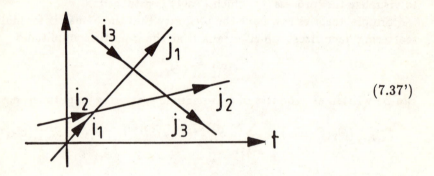

$$(7.37')$$

But (7.37') would have to be written as

$$S^{j_1 j_2 j_3}_{i_1 i_2 i_3}(\Theta_1, \Theta_2, \Theta_3) = \sum_{k_1 k_2 k_3} S^{k_2 k_3}_{i_2 i_3}(\Theta_{23}) S^{k_1 k_3}_{i_1 k_3}(\Theta_{13}) S^{j_1 j_2}_{k_1 k_2}(\Theta_{12}) \quad . \qquad (7.36')$$

So, as a consistency condition resulting from the (assumption of) the factorizability of the S-matrix, one obtains the equation

$$\sum_{k_1 k_2 k_3} S^{k_1 k_2}_{i_1 i_2}(\Theta_{12}) S^{j_1 k_3}_{k_1 i_3}(\Theta_{13}) S^{j_2 j_3}_{k_2 k_3}(\Theta_{23})$$

$$\overset{!}{=} \sum_{k_1 k_2 k_3} S^{k_2 k_3}_{i_2 i_3}(\Theta_{23}) S^{k_1 k_3}_{i_1 k_3}(\Theta_{13}) S^{j_1 j_2}_{k_1 k_2}(\Theta_{12}) \quad , \qquad (7.39)$$

which *is* (with $x = e^{\Theta}$, and introducing the tensor products according to the way of summing over indices) Eq. (7.32).

Notes and references

The method of classical r-matrices goes back to [1]. The QYBE had appeared first when solving the problem of N non-relativistic one-dimensional quantum mechanical particles interacting with each other by δ-function potentials [2]. Concerning its appearance (and importance) in statistical mechanics, see [3]. A reprint volume giving a good view on the development and diversity of the whole subject is [4]. Concerning relations to braid groups, the theory of knots and links and the exchange algebras of $1 + 1$ dimensional conformal field theories, see e.g. [5]. For a recent view on multi dimensional (and other) generalisations see [7]. Concerning generalized Lax pairs and the CYBE see e.g. [8]. A large class of solutions of the CYBE was obtained in [6]. For quantum R-matrices of Toda systems (including a classical r-matrix of the finite dimensional periodic Toda lattice that differs from the one given in (7.29)), see [10].

[1]   E.K. Sklyanin; J. Sov. Math. 19 (1982) 1546, LOMI preprint E-3-79 (1980).

[2]   C.N. Yang; Phys. Rev. Lett. 19 (1967) 1312
        J.B. Mc.Guire; J. Math. Phys. 5 (1964) 622.

[3]   R.J. Baxter; *Exactly solved Models in Statistical Mechanics*, Academic Press 1982.

[4]   M. Jimbo (ed.); *Yang Baxter Equation in Integrable Systems*, World Scientific 1989/90.

[5]   J. Fröhlich; Nucl. Phys. B 5B (1988) 110
        K.H. Rehren, B. Schroer; Nucl. Phys. 312 (1989) 715.

[6]   A.A. Belavin, V.G. Drinfeld; Funct. Anal. Appl. 16 (83) 159.

[7]   J.M. Maillet; *Integrable Systems and ...* CERN TH 5836/90.

[8]   M. Bordemann; Comm. Math. Phys. 135 (1990) 201.

[9]   A.B. Zamolodchikov, A.B. Zamolodchikov; Ann. Phys. 120 (1979) 253.

[10]   M. Jimbo; Comm. Math. Phys. 102 (1986) 537.

[11]   J. Avan, M. Talon; Nucl. Phys. B352 (1991) 215.

8 Integrable Systems and gl(∞)

Given some finite dimensional integrable system (with N degrees of freedom) it is natural to consider the limit $N \to \infty$. The transition from (quantum) mechanics to (quantum) field theory having many different aspects, and in concrete models various physical interpretations, we will choose a purely *algebraic* point of view – as in the context of integrable systems this turns out to be rather useful (yielding, e.g., a large class of possibilities which would have been difficult to reach by purely physical considerations).

Supposing that the equations of motion of some finite dimensional system are equivalent to a Lax-type equation,

$$\dot{L} = [L, M] \tag{8.1}$$

where L and M are finite dimensional matrices, one simply asks how one can maintain such an equation if L and M become infinite dimensional. Remembering that the proof of time-independence of

$$Q_k = \operatorname{Tr} L^k \tag{8.2}$$

only uses

$$\operatorname{Tr}[A, B] = 0 \tag{8.3}$$

and

$$[AB, C] = A[B, C] + [A, C]B \quad , \tag{8.4}$$

it is clear that all one has to require in the infinite dimensional case is the existence of a well defined 'invariant' trace operation (satisfying (8.3)) on the infinite dimensional Lie algebra in which L and M lie (and which has to have the additional property (8.4)).

If one considers L and M as infinite dimensional matrices, (8.3) and (8.4) are by no means trivially satisfied, as taking the product of two infinite dimensional matrices involves infinite summations – which might not converge. Of course, once an (associative) product $A \cdot B$ is well defined (as it is, e.g., for matrices possessing only a finite number of non-zero diagonals) (8.4) is automatically satisfied (and in the case of matrices, (8.3) then as well). As we will see, however, one may as well consider L and M as lying in some infinite dimensional Lie algebra (with trace) *without* underlying associative structure.

Before considering some specific examples (starting with the Toda chain) let us look at what kind of Lie algebras may arise as limits of finite dimensional matrix algebras, e.g. gl($N, \mathbb{C}$) (the Lie algebra of all complex $N \times N$ matrices).

The simplest basis of gl(N) consists of taking the N^2 matrices E_{ij} (each having all entries equal to 0, except the one at position ij, respectively, which is 1). They satify

$$[E_{ij}, E_{kl}] = \delta_{jk}E_{il} - \delta_{li}E_{kj} \quad \left(i, j, k, l = 1, \ldots, N\right) \quad . \tag{8.5}$$

Let us then consider an *arbitrary* basis $\{T_a\} = 1, 2, \ldots, N^2$ of gl(N), which by definition must be related to the E_{ij} basis by an invertible linear transformation Γ:

$$T_a = \gamma_{a,ij} E_{ij} \quad . \tag{8.6}$$

For the structure constants (of gl(N)) in the T_a basis one then finds

$$[T_a, T_b] = f_{ab}{}^c T_c \tag{8.7}$$

$$f_{ab}{}^c = \gamma_{a,ij}\gamma_{b,jk}(\Gamma^{-1})_{ik,c} - (a \leftrightarrow b) \quad a, b, c = 1, \ldots, N^2 \tag{8.8}$$

As $N \to \infty$ it may very well happen, that the structure constants (8.8) do have a finite limit $f_{ab}^{\infty c}$ (for each fixed triple (abc)) although (8.6) is ill–defined (due to numerical divergence of the $\gamma_{a,ij}$, or the sum over i and j) or defines infinite dimensional matrices which can not be multiplied. In this case (and under some mild assumptions on the $f_{ab}^{\infty c}$ to assure validity of the Jacobi identity)

$$[T_a, T_b] = f_{ab}^{\infty c} T_c \quad a, b, c = 1, \ldots, \infty \tag{8.9}$$

will define an infinite dimensional Lie algebra L, which can be viewed as a gl($N \to \infty$) limit although it will in general *not* be isomorphic to the 'standard' gl(∞), which is defined as consisting of all infinite dimensional matrices with only finitely many non zero elements, i.e. all *finite* linear combinations of basis-elements E_{ij},

$$[E_{ij}, E_{kl}] = \delta_{jk}E_{il} - \delta_{li}E_{kj} \quad i, j, k, l = 1, 2, \ldots, \infty \quad . \tag{8.10}$$

As we will shortly use such (non-isomorphic) gl($N \to \infty$) limits to guess infinite dimensional analogues of the finite Toda chain, let us consider in detail an example of the above situation. Let

$$\gamma_{\mathbf{m},ij} = \frac{iN}{4\pi M}\omega^{\frac{1}{2}m_1 m_2 + (i-1)m_1}\delta_{i+m_2, j\,\mathrm{mod}\,N)}$$

$$\mathbf{m} = (m_1, m_2) \quad m_1, m_2 = -\frac{N-1}{2}, \ldots, +\frac{N-1}{2} \tag{8.11}$$

$$M, N \in \mathbb{N} \quad N \text{ odd} \quad \omega = e^{4\pi i \frac{M}{N}} \quad .$$

One can then check, that

$$(\Gamma^{-1})_{ij,\mathbf{m}} = -\frac{4\pi iM}{N^2}\omega^{\frac{1}{2}m_1 m_2 + m_1(1-j)}\delta_{m_2, j-i(\mathrm{mod}\,N)} \tag{8.12}$$

and that, according to (8.8), the structure constants in the $\{T_\mathbf{m}\}$ basis will be

$$f_{\mathbf{mn}}^{\mathbf{k}} = \frac{N}{2\pi M}\sin\left(\frac{2\pi M}{N}(\mathbf{m} \times \mathbf{n})\right)\delta_{\mathbf{m+n,k}\ (\mathrm{mod}\,N)} \quad . \tag{8.13}$$

Writing out the $T_\mathbf{m}$ basis as $N \times N$ matrices one finds that they are conveniently written as

$$T_\mathbf{m} = \frac{iN}{4\pi M} \omega^{\frac{1}{2} m_1 m_2} g^{m_1} h^{m_2} \quad,$$

$$g = \begin{pmatrix} 1 & 0 & 0 & \cdots & 0 \\ 0 & \omega & 0 & \cdots & 0 \\ 0 & 0 & \omega^2 & \ddots & \vdots \\ \vdots & & \ddots & \ddots & 0 \\ 0 & \cdots & & 0 & \omega^{N-1} \end{pmatrix} \quad , \quad h = \begin{pmatrix} 0 & 1 & 0 & \cdots & 0 \\ 0 & 0 & 1 & \ddots & \vdots \\ \vdots & \ddots & \ddots & \ddots & 0 \\ 0 & 0 & \cdots & 0 & 1 \\ 1 & 0 & \cdots & 0 & 0 \end{pmatrix} \quad , \tag{8.14}$$

$$hg = \omega gh \quad .$$

Obviously, neither (8.14) nor (8.11) makes sense if $N = \infty$. Eq. (8.13), however, converges to 'structure constants'

$$f_{\mathbf{mn}}^{\infty \, \mathbf{k}} = (\mathbf{m} \times \mathbf{k}) \, \delta_{\mathbf{m+n,k}} \qquad \mathbf{k}, \mathbf{m}, \mathbf{n} \in \mathbb{Z}^2 \tag{8.15}$$

(which trivially satisfy the Jacobi identity). So one is led to an infinite dimensional Lie algebra L_0, defined as the linear span of basis elements $T_\mathbf{m}$ ($\mathbf{m} \in \mathbb{Z}^2$), and commutation relations

$$[T_\mathbf{m}, T_\mathbf{n}] = (\mathbf{m} \times \mathbf{n}) \, T_{\mathbf{m+n}} \qquad \mathbf{m}, \mathbf{n} \in \mathbb{Z}^2 \quad . \tag{8.16}$$

L_0 (as defined via (8.16)) is well known, and coincides (up to completion) with the Poisson algebra $(\text{diff}_A T^2)_\mathbb{C}$ of smooth (complex valued) functions on the torus, with Poisson bracket

$$[f(\varphi_1, \varphi_2), g(\varphi_1, \varphi_2)] = i\epsilon_{rs} \partial_r f \partial_s g \tag{8.17}$$

and canonical basis

$$T_\mathbf{m} = ie^{im\varphi} \quad . \tag{8.18}$$

Moreover, one can use (8.11) to obtain a whole (infinite) class of (non–isomorphic) infinite dimensional Lie algebras, by choosing $M = M(N)$ in (8.11) such that

$$\frac{M(N)}{N} \to \Lambda \quad \text{irrational} \quad . \tag{8.19}$$

Then (8.13) will converge to

$$f_{\mathbf{mn}}^\mathbf{k} = \frac{1}{2\pi\Lambda} \sin(2\pi\Lambda(\mathbf{m} \times \mathbf{n})) \, \delta_{\mathbf{m+n,k}} \tag{8.20}$$

(again satisfying the Jacobi identity). Let us call the Lie algebra corresponding to (8.20) L_Λ,

$$[T_\mathbf{m}, T_\mathbf{n}] = \frac{1}{2\pi\Lambda} \sin(2\pi\Lambda(\mathbf{m} \times \mathbf{n})) \, T_{\mathbf{m+n}} \quad . \tag{8.21}$$

One finds [5] that for Λ irrational, $\Lambda \in (0, 1/4)$, all these Lie algebras are different (non-isomorphic), and that for $\Lambda = p/q$ rational, L_Λ contains an ideal of finite codimension, which (when divided out) yields gl($q, \mathbb{C}$).

In the next lecture we will also use that (8.21) has the following representation [3] in terms of operators:

$$T_{\mathbf{m}} = \frac{i}{4\pi\Lambda} e^{i\mathbf{m}\cdot\mathbf{A}} \quad , \quad [A_1, A_2] = 4\pi i\Lambda \cdot \mathbb{1} \quad , \tag{8.22}$$

e.g. $A_1 = \theta \in [0, 2\pi]$, $A_2 = -4\pi i\Lambda \partial/\partial\theta$.

Notes and references

The formulae are mostly from [1], [2]. L_A (Eq. (8.21)) was introduced in [4]. The mutual non–isomorphy of the L_A for different irrational values of $A \in (0, 1/4)$ was proven in [5]. An embedding of L_A into the Lie algebra $gl_J(\infty)$ of infinite dimensional matrices with finitely many non–zero diagonals was given in [6]. L_0 was studied in detail in [7]. The convergence of structure constants in a non canonical basis leading to non trivial $gl(N \to \infty)$ limits goes back to [8]. The idea to use these non obvious $gl(N \to \infty)$ limits in the context of integrable systems seems to be new, and was used in [9]. From a more abstract point of view, algebras L_A, and related integrable systems, were also considered in [10].

[1]   J. Hoppe; Int. J. of Mod. Phys. A, Vol. 4 #19 (1989) 5235.
[2]   M. Bordemann, J. Hoppe, P. Schaller, M. Schlichenmaier;
       Comm. Math. Phys. 138 #2 (1991) 209.
[3]   J. Hoppe; Phys. Lett. B 215 (1988) 706.
[4]   D.B. Fairlie, P. Fletcher, C.K. Zachos; Phys. Lett. B 218 (1989) 203.
[5]   J. Hoppe, P. Schaller; Phys. Lett. B 237 (1990) 407.
[6]   E.G. Floratos; Phys. Lett. B 232 (1989) 467.
[7]   V. Arnold; Ann. Institut Fourier XVI (1966) 319.
       Mathematical Methods of Classical Mechanics, Springer Verlag, New York, 1978.
[8]   J. Goldstone; unpublished.
       J. Hoppe; MIT Ph.D. Thesis 1982.
[9]   M. Bordemann, J. Hoppe, S. Theisen; Phys. Lett. B (1991) 374.
[10]   M.V. Saveliev, A.M. Vershik, Comm. Math. Phys. 126 (1989) 367,
       Phys. Lett. A 143 (1990) 121.

9 Infinite Dimensional Toda Systems

Let us now use $N \to \infty$ limits of $\mathrm{gl}(N, \mathbb{C})$ in the context of a concrete physical model, the periodic Toda chain:

$$H_N = \frac{1}{2m} \sum_{i=1}^{N} p_i^2 + g^2 \sum_{i=1}^{N} e^{2b(q_i - q_{i+1})} \tag{9.1}$$

$$L^{(N)} = \begin{pmatrix} p_1 & a_1 & 0 & \cdots & & 0 & a_N \\ a_1 & p_2 & a_2 & \ddots & & 0 & 0 \\ 0 & a_2 & p_3 & \ddots & & \ddots & \vdots \\ \vdots & \ddots & \ddots & \ddots & & a_{N-2} & 0 \\ 0 & \cdots & 0 & a_{N-2} & & p_{N-1} & a_{N-1} \\ a_N & 0 & \cdots & & 0 & a_{N-1} & p_N \end{pmatrix} .$$

Physically, one could use a picture where $q_i(t) = q(\varphi_i, t)$ is the value of some $1+1$ dimensional field at the spatial point φ_i (let us drop t, for the moment). Letting $\varphi_{i+1} = \varphi_i + \epsilon$, $\nu = 1/\epsilon$, $g^2 = \gamma \epsilon$ and $\epsilon = 2\pi/N \to 0$ (as $N \to \infty$), one could interpret the potential energy in (9.1) as approaching

$$\gamma \int_0^{2\pi} e^{2q'(\varphi)} d\varphi \quad , \tag{9.3}$$

where $' = \partial/\partial\varphi$, and q has to be a periodic function of φ, $q(\varphi_i + 2\pi) = q(\varphi_i)$, due to the fact that we started from the *periodic* lattice (i.e. $q_{N+1} = q_1$ in (9.1)).

Instead of looking at the other conserved charges $(Q_{k>2})$ in the limit $N \to \infty$ (this discussion is rather delicate) let us use our knowledge about $\mathrm{gl}(N \to \infty)$ to guess directly a Lax pair for the infinite dimensional limit (9.3) of (9.1) – obviously, the kinetic energy of (9.1) will lead to

$$\frac{1}{2\mu} \int_0^{2\pi} p^2(\varphi) d\varphi \tag{9.3'}$$

when $m = \mu/2\pi N = \mu\epsilon$ is properly chosen.

Let us assume that $L^{(\infty)}$ and $M^{(\infty)}$ will lie in the Lie Poisson algebra L_0 of complex functions on the torus, $(\mathrm{diff}_A T^2)_C$. Comparing (9.14) with (9.18) one sees that, more or less, $e^{i\varphi_1}$ can be regarded as the analogue of the finite dimensional matrix $g = \mathrm{diag}(1, \omega, \omega^2, \ldots, \omega^{N-1})$, $\omega = e^{4\pi i M/N}$, while the ('shift'–) matrix h 'becomes' $e^{\pm i\varphi_2}$. As (9.2) is of the form

$$L^{(N)} = P + \frac{1}{2}\left(Ah + h^{-1}A\right) \quad,$$

$$A = 2\operatorname{diag}(a_1, a_2, \ldots, a_N) = A[\mathbf{q}] \quad P = \operatorname{diag}(p_1, \ldots, p_N) \quad, \tag{9.4}$$

one may expect

$$L^{(\infty)} = L(\varphi_1, \varphi_2) = p(\varphi_1) + \frac{1}{2}\left(F[\mathbf{q}]e^{i\varphi_2} + e^{-i\varphi_2}F[\mathbf{q}]\right)$$

$$= p(\varphi_1) + \cos\varphi_2 \, F[q] \tag{9.5}$$

— note that the second line in (9.5) is trivially obtained from the first one as in contrast with the finite dimensional case, where $[A, h] \neq 0$, $e^{i\varphi_2}$ and $F[q]$ commute. In fact, already knowing that the limit of a_i will be proportional to $e^{q'(\varphi_1)}$, one could already guess that

$$F[q] = F[q'] = e^{q'(\varphi)} \quad. \tag{9.6}$$

In any case, M should be of the form

$$M = c\left(F[q]e^{i\varphi_2} - e^{-i\varphi_2}F[q]\right)$$

$$= 2ic\sin\varphi_2 F[q] \quad. \tag{9.7}$$

Indeed, the Lax equation in the Lie algebra L_0,

$$\dot{L} = [L, M] := i\left(\frac{\partial L}{\partial \varphi_1}\frac{\partial M}{\partial \varphi_2} - \frac{\partial L}{\partial \varphi_2}\frac{\partial M}{\partial \varphi_1}\right) \tag{9.8}$$

with L and M as in (9.5) and (9.7), and $c = -1/2$, will be consistent, provided F is chosen as in (9.6), and the equations of motion

$$\dot{q} = p \quad, \quad \dot{p} = q'' e^{2q'(\varphi_1)} \tag{9.9}$$

hold — which *are* the equations of motion for

$$H = \frac{1}{2}\int_0^{2\pi}\frac{d\varphi}{2\pi}\left(p^2(\varphi) + \frac{1}{2}e^{2q'(\varphi)}\right) \tag{9.10}$$

— cp. (9.2),(9.2'). In order to deduce conservation laws from the Lax equation (9.8), we need an invariant trace (cp. (9.3)) and the derivation property (cp. (9.4)). Fortunately, both hold in L_0 (although the Poisson bracket in L_0 does *not* come from an underlying associative structure), if we take

$$\operatorname{Tr} f := \int_0^{2\pi}\frac{d\varphi_1}{2\pi}\int_0^{2\pi}\frac{d\varphi_2}{2\pi}f(q_1, q_2) \quad. \tag{9.11}$$

So

$$Q_k = \operatorname{Tr} L^k = \int\int\frac{d\varphi_1 d\varphi_2}{4\pi^2}L^k(\varphi_1\varphi_2) \quad. \tag{9.12}$$

Q_k will be time independent, due to (9.8). Note that $H = \frac{1}{2}Q_2$ (the integral over φ_2 eliminates the term linear in p and results in numerical factors for p^2

and $\cos^2 \varphi_2 F^2$), and that the associative multiplication used to define L^k is the ordinary commutative multiplication

$$(f \cdot g)(\varphi_1, \varphi_2) = f(\varphi_1, \varphi_2) \cdot g(\varphi_1, \varphi_2) \quad . \tag{9.13}$$

Later we will prove that that the conserved charges (9.12) are in involution,

$$\{Q_m, Q_n\}_{\text{P.B.}} := \int_0^{2\pi} \frac{d\varphi}{2\pi} \left(\frac{\partial Q_m}{\partial q(\varphi)} \frac{\partial Q_n}{\partial p(\varphi)} - \frac{\partial Q_m}{\partial p(\varphi)} \frac{\partial Q_n}{\partial q(\varphi)} \right) = 0 \quad \forall m, n \in \mathbb{N} \quad . \tag{9.14}$$

These results can be generalized in two different directions: First of all, due to the fact that elements in L_0 are commutative after application of the Lie-bracket, i.e.

$$f\{g, h\} = \{g, h\}f \quad ,$$

one may hope that one can obtain more integrable systems from L_0 than there are finite dimensional ones known (See 10).

Secondly, let us try to use the $\text{gl}(N \to \infty)$ limits L_A, with representation (cp. (9.22))

$$T_{\mathbf{m}} = \frac{i}{4\pi\Lambda} \omega^{\frac{1}{2}m_1 m_2} e^{im_1\theta} e^{m_2\lambda\partial_\theta} \tag{9.15}$$

$$\lambda = 4\pi\Lambda \quad , \quad \omega = e^{i\lambda} \quad , \quad \partial_\theta = \frac{\partial}{\partial\theta}$$

for the Toda system. Comparing (9.15) with the form of the generators $T_{\mathbf{m}}$ of $\text{gl}(N)$, (9.14), one sees a correspondence

$$g \leftrightarrow e^{i\theta} \quad , \quad h \leftrightarrow e^{\lambda\partial_\theta} \quad . \tag{9.16}$$

So, in analogy to (9.4), one may try the Ansatz

$$L = p(\theta) + f(\theta) \cdot T + T^{-1} \cdot f(\theta) \qquad T = e^{\lambda\partial_\theta} \quad . \tag{9.17}$$
$$M = f(\theta) \cdot T - T^{-1} \cdot f(\theta)$$

Calculating $\dot{L} - [L, M]$, one finds

$$\begin{aligned}
&\dot{p} + 2\left(f^2(\theta) - f^2(\theta - \lambda)\right) \\
&+ \left(\dot{f} + f\left(p(\theta + \lambda) - p(\theta)\right)\right) \cdot T \\
&+ T^{-1} \cdot \left(\dot{f} + f\left(p(\theta + \lambda) - p(\theta)\right)\right)
\end{aligned} \tag{9.18}$$

where one has used that for any function $g(\theta)$

$$T \cdot g(\theta) = g(\theta + \lambda) \cdot T \quad . \tag{9.19}$$

As the coefficients of the three different powers of T must be independently zero, one finds

$$f(\theta) = a \cdot e^{b(q(\theta) - q(\theta + \lambda))} \quad , \tag{9.20}$$

$$p(\theta) = b\dot{q}(\theta) \quad , \quad \dot{p} = 2a^2 \left(e^{2b(q(\theta-\lambda)-q(\theta))} - e^{2b(q(\theta)-q(\theta+\lambda))} \right) \quad . \tag{9.21}$$

So the Ansatz (9.17) is consistent, and the Lax equation in the Lie Algebra L_Λ (actually, its partial completion, if we allow $f(\theta)$ to have infinitely many non zero Fourier coefficients),

$$\dot{L} = [L, M]_\Lambda \tag{9.22}$$

is equivalent to the equations of motion corresponding to the Hamiltonian

$$H = \frac{1}{2b} \int_0^{2\pi} \frac{d\theta}{2\pi} \left\{ p^2(\theta) + a^2 \left(e^{2b(q(\theta)-q(\theta+\lambda))} + e^{2b(q(\theta-\lambda)-q(\theta))} \right) \right\} \quad . \tag{9.23}$$

Can we define an invariant trace on L_Λ? Writing a general element $F \in L_\Lambda$ as

$$F = \sum_m f_m(\theta) T^m \tag{9.24}$$

it is not difficult to prove that

$$\widehat{\mathrm{Tr}} F := \int_0^{2\pi} \frac{d\theta}{2\pi} f_0(\theta) \tag{9.25}$$

will be invariant,

$$\widehat{\mathrm{Tr}}(F \cdot G) = \widehat{\mathrm{Tr}}(G \cdot F) \quad . \tag{9.26}$$

So

$$Q_k := \widehat{\mathrm{Tr}} L^k \tag{9.27}$$

will be time independent, due to (9.22). Eq. (9.17) and Eq. (9.22) can be lifted to describe a $2+1$ dimensional field theory:

$$\begin{aligned} L &= p + fT + T^{-1}f \\ M &= fT - T^{-1}f + m(x, \theta, t) \end{aligned} \tag{9.28}$$

will satisfy the 'zero-curvature condition'

$$[\partial_t + M, \partial_x + L] = 0 \tag{9.29}$$

provided $m = b\partial_x q(\theta, x, t)$ and

$$\ddot{q} - \partial_x^2 q = \frac{2a^2}{b} \left(e^{2b(q(\theta-\lambda)-q(\theta))} - e^{2b(q(\theta)-q(\theta+\lambda))} \right) \quad . \tag{9.30}$$

Note that choosing $b = -1/\lambda$, $a = 1/2\lambda$, and q x–independent, (9.30) (as well as (9.21)) reduces to (9.9) as $\lambda \to 0$.

Finally, one should mention that (after some change of variables, $x_\pm = x \pm t$) the equations of motion (9.29) can also be obtained from a somewhat simpler, unsymmetric form of L and M,

$$\begin{aligned} L &= T + p(\theta, x_\pm) \\ M &= T^{-1} f(\theta, x_\pm) \quad , \end{aligned} \tag{9.31}$$

which is the starting point for a whole hierarchy of integrable systems [4].

Notes and references

Eq. (9.10) is a particular example of the systems discussed in [1], and coincides with a reduction of a system considered in [2]. The discussion of the corresponding L_A systems is based on [5]. Eq. (9.29) has also been considered in [3]. For a discussion of the Toda-hierarchy and corresponding τ-functions, see e.g. [4].

[1]   M. Bordemann, J. Hoppe, S. Theisen; Phys. Lett. B (1991) 374.
[2]   M.V. Saveliev, A.M. Vershik; Comm. Math. Phys. 126 (1989) 367,
        Phys. Lett. A 143 (1990) 121.
[3]   A. Degasperis, D. Lebedev, M. Olshanetsky, S. Pakuliak, A. Perolomov, P. Santini;
        Bonn University Preprint HE–90–14.
[4]   K. Ueno, K. Takasaki; Advanced Studies in Pure Mathematics 4 (1984) 1.
[5]   J. Hoppe, M. Olshanetsky, S. Theisen; *Dynamical Systems on Quantum Tori Algebras*, KA-THEP-10-91.

10 Integrable Field Theories from Poisson Algebras

Usually, the Lie product $[L, M]$ is realized as the antisymmetric part of some non-commutative, associative multiplication $\cdot$,

$$[L, M] = L \cdot M - M \cdot L \quad . \tag{10.1}$$

In the case of $[\,,\,]$ being a Poisson-bracket $\{\,,\,\}$ on some space of functions from a (compact) manifold to the field $\mathbb{C}$ (or $\mathbb{R}$), e.g.

$$L, M : T^2 \to \mathbb{C}$$

$$\{L, M\} := \frac{\partial L}{\partial \varphi_1} \frac{\partial M}{\partial \varphi_2} - \frac{\partial L}{\partial \varphi_2} \frac{\partial M}{\partial \varphi_1} \tag{10.2}$$

— which is *not* of the type (10.1) — there exists an additional *commutative* associative multiplication, which is the one implied when writing L^k, e.g., and which is the ordinary multiplication of functions. Consequently,

$$Q_k^{(\infty)} = \operatorname{Tr} L^k = \int \frac{d\varphi_1 d\varphi_2}{(2\pi)^2} \left(L\left(\varphi_1, \varphi_2\right)\right)^k = \int d\Omega \left(L\left(\Omega\right)\right)^k \tag{10.3}$$

is 'simpler' then e.g. the matrix trace

$$\operatorname{Tr} L^k = \sum_{i=1}^{N} \left(L^k \right)_{ii} = \sum_{i=1}^{N} \sum_{i_2 \ldots i_k} L_{i i_2} L_{i_2 i_3} \cdots L_{i_k i} = Q_k^{(N)} \quad . \tag{10.4}$$

One could say that (10.4) is 'non-local', in contrast with (10.3). Moreover (concerning the practical point of view) it is rather obvious that the involutivity of some set of conserved charges will be simpler to prove in the infinite dimensional case (10.3), than in the finite dimensional case, (10.4).

In any case, one can speculate that the simplifying features of the Poisson algebra $N \to \infty$ limit of $\mathrm{gl}(N)$ will allow one to find integrable systems whose interaction is of a more general type than those known to be integrable for finite dimensional models $(1/x^2, e^x, \ldots)$.

Guided by (9.5), where the tridiagonal nature of the finite dimensional Toda Lax matrix L was converted into an Ansatz containing only the three harmonics $e^{\pm i \varphi_2}$ and $1 = e^0$ (the latter resulting from diagonal matrices), one could try the more general Ansatz

$$L(\varphi_1, \varphi_2) = p(\varphi_1) + e(\varphi_2) F[q(\varphi_1)]$$
$$M(\varphi_1, \varphi_2) = -f(\varphi_2) G[q(\varphi_1)] \tag{10.5}$$

(the illogical minus sign taken from [1]), where e and f are now allowed to contain an arbitrary number of harmonics $e^{\pm im\varphi_2}$ (corresponding to finite dimensional matrices with no restriction on the number of non–zero diagonals); physically, this includes arbitrary pair potentials, in contrast to considering only nearest neighbour interactions). Calculating both sides of

$$\dot{L} = \{L, M\} \qquad (10.6)$$

one easily finds that F must be a *function* of $q' = \partial q/\partial\varphi_1$, G must equal F' (here, and below, ' always denotes differentiation with respect to the relevant argument, not counting t; thus $F'(q') = \partial F(q')/\partial q'$) and that F, e and f must satisfy

$$FF'' = \alpha F'^2 \quad (\alpha \in \mathbb{R}) \quad , \qquad (10.7)$$

$$e = -f' \quad , \quad f'^2 - \alpha ff'' = \kappa \quad (\kappa \in \mathbb{R}) \qquad (10.8)$$

for (10.6) to be consistent, and to be equivalent to a pair of Hamiltonian equations of motion, which then are

$$\dot{q} = p \quad , \quad \dot{p} = \kappa q'' F'^2(q') \quad . \qquad (10.9)$$

The corresponding Hamiltonian,

$$H = \frac{1}{2} \int_0^{2\pi} \frac{d\varphi}{2\pi} \left(p^2(\varphi) + \lambda F^2 \right) \quad , \quad \lambda = \kappa(\alpha + 1) \qquad (10.10)$$

is of course nothing but $\frac{1}{2} \operatorname{Tr} L^2$. While (10.7) can be solved explicitely,

$$F = ae^{bq'} \quad (\alpha = 1)$$
$$F = (a + bq')^{\frac{1}{1-\alpha}} \quad (\alpha \neq 1) \quad , \qquad (10.11)$$

explicit solutions of (10.8) are known only for special values of α, notably $\alpha = 1$ ($\rightarrow$ Toda), $\alpha = 2$ ($\rightarrow$ inverse square potential), and $\alpha = \frac{1}{2}$ ($\rightarrow$ quartic interaction). Nevertheless, smooth periodic solutions of (10.8) do exist for a continuous range of α (at least for $0 < \alpha \leq 1$), and it turns out that in these cases one can explicitly write down the conserved charges and prove that they are in involution (!), on the basis of merely noting that (10.8) implies that

$$\kappa_n := \int_0^{2\pi} \frac{d\varphi_2}{2\pi} e^n = \begin{cases} 0 & \text{if } n \text{ is odd} \\ \kappa^{\frac{n}{2}} I_n(\alpha), & n \text{ even} \end{cases} \quad , \qquad (10.12)$$

$$I_{2k} = \frac{(2k-1)(2k-3)\cdots 1}{(2k-1+\alpha)(2k-3+\alpha)\cdots(\alpha+1)} \quad . \qquad (10.13)$$

(10.12)/(10.13) can easily be proved by partial integration (using (10.8)) and solving the resulting recursion formula for I_k,

$$I_n = \frac{n-1}{n-1+\alpha} I_{n-2} = \cdots \quad ; \quad I_{2k+1} = 0 \text{ as } I_1 = 0 \quad .$$

So, without knowing e explicitly, we can write down the conserved charges $Q_n = \operatorname{Tr} L^n$ in '1-dimensional' form:

$$Q_n = \operatorname{Tr} L^n = \sum_{k=0}^{n} \binom{n}{k} \kappa_k \int \frac{d\varphi_1}{2\pi} \left(p^{n-k} F^k \right) \quad , \tag{10.14}$$

the κ_k appearing as numerical 'coupling' constants.

When calculating the Poisson bracket of Q_n with Q_m,

$$\{Q_n, Q_m\} = \int_0^{2\pi} \frac{d\varphi}{2\pi} \left(\frac{\delta Q_n}{\delta q(\varphi)} \frac{\delta Q_m}{\delta p(\varphi)} - \frac{\delta Q_n}{\delta p(\varphi)} \frac{\delta Q_m}{\delta q(\varphi)} \right) \quad , \tag{10.15}$$

one is led to rather involved combinatorical identities — implying that (10.15) is identically zero for all m and n —,

$$\sum_{k=0}^{\left[\frac{m}{2}\right]} \binom{m}{2k} \binom{n}{2J - 2k} I_{2k}(\alpha) I_{2J-2k}(\alpha) \cdot$$

$$\cdot \left\{ (m - 2k)(m - 2k - 1)(J - k)(2J - 2k + \alpha - 1) \right. \tag{10.16}$$

$$\left. - (n - 2J + 2k)(n - 2J + 2k - 1)k(2k + \alpha - 1) \right\} = 0$$

$$J = 1, 2, \ldots, \left[\frac{m+n}{2}\right] - 1 \quad ,$$

which should also be useful when quantizing these theories.

The easiest way, however, to prove involutivity of the conserved charges Q_n is to go back to their definition (10.3), and note that (10.15) will be zero (for all m and n) if the generating functional

$$Q(\lambda) := \int \frac{d\varphi_1 d\varphi_2}{(2\pi)^2} e^{\lambda L(\varphi)} \tag{10.17}$$

commutes with itself at different values of $\lambda \in \mathbb{R}$,

$$\{Q(\lambda), Q(\mu)\} = \int \frac{d\varphi \, d\varphi'}{(2\pi)^4} \left\{ e^{\lambda L(\varphi)}, e^{\mu L(\varphi')} \right\} = 0 \quad \forall \lambda, \mu \quad . \tag{10.18}$$

When calculating the l.h.s. of (10.18) one finds (after several integrations by parts) that (10.7) and (10.8) (together with periodicity of f, or f vanishing at 0 and 2π) are sufficient to prove (10.18):

$$\frac{1}{\lambda\mu}\int\left\{e^{\lambda L(\varphi)},e^{\mu L(\varphi')}\right\}$$

$$=\int\frac{d\varphi\,dy\,d\varphi'\,dy'}{(2\pi)^4}\left\{p(\varphi)+e(y)F_\varphi,p(\varphi')+e(y')F_{\varphi'}\right\}$$

$$L(\varphi)=p(\varphi)+e(y)F(q'(\varphi))=:L_{\varphi y}$$

$$=\int\left(e(y)F'_\varphi\partial_\varphi\delta(\varphi-\varphi')-e(y')F'_{\varphi'}\partial_{\varphi'}\delta(\varphi-\varphi')\right)e^{\lambda L_{\varphi y}+\mu L_{\varphi'y'}}$$

$$\frac{\delta p(\varphi)}{\delta p(\varphi'')}=\delta(\varphi-\varphi'')\quad,\quad\frac{\delta q'(\varphi)}{\delta q(\varphi'')}=\partial_\varphi\delta(\varphi-\varphi'')=-\partial_{\varphi''}\delta(\varphi-\varphi'')$$

$$=\int\frac{d\varphi\,dy\,dy'}{(2\pi)^3}\left(\mu e(y)F'_\varphi\left(p'+e(y')F'q''\right)-\lambda e(y')F'\left(p'+e(y)F'q''\right)\right)e^{\lambda L_y+\mu L_{y'}}$$

partial integration and $\displaystyle\int\frac{d\varphi'}{2\pi}$

$$=\int\left[(\mu-\lambda)e(y)e(y')F'^2q''e^{\lambda L_y}e^{\mu L_{y'}}\right.$$

$$+\frac{\mu}{\lambda+\mu}e(y)\left(e^{(\lambda+\mu)p}\right)'F'e^{F(\lambda e_y+\mu e'_y)}$$

$$\left.-\frac{\lambda}{\lambda+\mu}e(y')\left(e^{(\lambda+\mu)p}\right)'F'e^{F(\lambda e_y+\mu e_{y'})}\right]$$

by partial integration:

$$=\int\left[(\mu-\lambda)e(y)e(y')F'^2e^{F(\lambda e_y+\mu e_{y'})}q''e^{(\lambda+\mu)p}\right.$$

$$-\frac{\mu}{\lambda+\mu}e(y)\left(F''+F'^2\left(\lambda e_y+\mu e_{y'}\right)\right)e^{F(\cdots)}q''e^{(\lambda+\mu)p}$$

$$\left.+\frac{\lambda}{\lambda+\mu}e(y')\left(F''+F'^2\left(\lambda e_y+\mu e_{y'}\right)\right)e^{F(\cdots)}q''e^{(\lambda+\mu)p}\right]$$

$$(\mu-\lambda)+\frac{\lambda^2-\mu^2}{\lambda+\mu}=0$$

$$=\frac{\lambda\mu}{\lambda+\mu}\int\left(e(y')^2-e(y)^2\right)F'^2q''e^{\lambda L_y+\mu L_{y'}}$$

$$+\frac{1}{\lambda+\mu}\underbrace{\int\left(\lambda e(y')-\mu e(y)\right)F''q''e^{\lambda L_y+\mu L_{y'}}}$$

with $e=-f'$, partial integration

$$=-\lambda\mu\int\left(ff''(y')-ff''(y)\right)FF''q''e^{\lambda L_y+\mu L_{y'}}$$

$$ff''=\frac{1}{\alpha}\left(f'^2-\kappa\right)$$

$$=\frac{\lambda\mu}{\lambda+\mu}\int\left(f'^2(y')-f'^2(y)\right)\underbrace{\left(F'^2-\frac{FF''}{\alpha}\right)}_{=0}q''e^{\lambda L_y+\mu L'_y}\quad=\quad0$$

Notes and references

These results were obtained in collaboration with M.Bordemann and S.Theisen [1]. For 2+1 dimensional generalizations, see 11. A naive (= a priori incorrect) quantum mechanical calculation (treating q and p as operators, $[q(\varphi), p(\varphi')] = i\hbar\delta(\varphi - \varphi')$, $[q''(\varphi), p(\varphi')] = i\hbar\delta''(\varphi - \varphi') \neq 0$, but q' and p as commuting — thus assuming the existence of a reguralisation in which $\delta'(0)$ can be replaced by 0) indicates that (10.16) might be used to prove commutativity of (to be properly defined) quantum charges Q_n.

[1]   M. Bordemann, J. Hoppe, S. Theisen; Phys. Lett. B 267 (1991) 374.

11 Generalized Garnier Systems and Membranes

Apart from exponential interactions and those proportional to the inverse square distance (as well as their generalisations to elliptic $\wp$–functions) there exists another type of non–linear interaction, which is rather special: the class of quartic potentials. In the finite dimensional case, FWM [1] used the Lax–representation of the non–linear Schrödinger model to generalize the original work of Garnier [2]. In a compact notation (and dropping a term quadratic in the coordinates, for simplicity) the FWM Hamiltonian can be written as

$$H_{FWM} = \frac{1}{2}\,\mathrm{Tr}(P^+ P + Q^+ Q Q^+ Q) \quad , \tag{11.1}$$

where P and Q are complex rectangular matrices containing the coordinate and momentum degrees of freedom, respectively,

$$Q = (q_{ia})_{\substack{i=1\ldots q \\ a=1\ldots p}} \qquad P = (p_{jb}) \tag{11.2}$$

$$\{q_{ia}, p_{jb}^*\} = \delta_{ij}\delta_{ab} \quad . \tag{11.3}$$

Equation (11.3) is a way of stating that decomposing Q and P into their real and imaginary parts, $Q = \frac{1}{\sqrt{2}}(Q^{(1)} + iQ^{(2)})$ and $P = \frac{1}{\sqrt{2}}(P^{(1)} + iP(2))$, the (real) matrix elements of $P^{(j)}$ are canonically conjugate to those of $Q^{(j)}$ $(j = 1, 2)$. The phase space is $4pq$-dimensional (i.e. twice the number of real matrix elements of some typical matrix). Note that this crucially differs from the previously considered finite dimensional systems where the degrees of freedom where more or less equal to the *square-root* of the number of matrix elements, i.e. equal to the number of *diagonal* elements. For a 'matrix model' like (11.1) it means that if there exists a Lax matrix it can not be a merely phase–space dependent matrix of the same dimension as Q and P (or, to make L rectangular, of size $N = p+q$), as in that case it woud not lead to sufficiently many conserved quantities (– due to the fact that there are only N independent trace–polynominals for any fixed $N \times N$ matrix). But if L contains a genuine spectral parameter ('genuine' meaning that $\mathrm{Tr}\,L^k(\lambda)$ will depend non-trivially on λ) the above argument is altered as in each order (k) the λ–dependence generates additional conserved quantities.

In any case, there exists the following Lax–representention for (11.1)

$$L = \lambda^2 A + \lambda \begin{pmatrix} 0 & -Q^+ \\ Q & 0 \end{pmatrix} + \frac{1}{a}\begin{pmatrix} -Q^+ Q & P^+ \\ P & QQ^+ \end{pmatrix}$$

$$M = -\lambda A - \begin{pmatrix} 0 & -Q^+ \\ Q & 0 \end{pmatrix} \quad ; \quad A = i\begin{pmatrix} -q\cdot \mathbb{1}_{p\times p} & 0 \\ 0 & p\cdot \mathbb{1}_{q\times q} \end{pmatrix} \tag{11.4}$$

$$\dot{L} = [L, M] \qquad a = i(p + q) = iN \quad ,$$

leading to conserved quantities

$$Q_{kl} := \left(\frac{\partial^l}{\partial \lambda^l} \operatorname{Tr} L^k(\lambda) \right)_{\lambda=0}^{\substack{k=1\ldots N=p+q \\ l=1\ldots 2k}} . \tag{11.5}$$

Part of the $N(N+1)$ 'charges' Q_{kl} are identically zero (or numerical, i.e. phase space and time independent constants), but sufficiently many (at least $N^2/2 \geq 2pq$) remain; that they are also in involution was shown in [3] and [4], where a classical r–matrix was obtained.

The existence of (11.4) is related to that of the 'hermitean symmetric space' [1], [5]

$$su(p+q)/s(u(p) \otimes u(q)) . \tag{11.6}$$

The matrix A spans the center of $s(u(p) \otimes u(q))$, and acts as multiplication by 0 or $\pm a$ upon commutation with some other matrix. Also note that the first two terms in L do not contribute to $[L, M]$, as (together) they are proportional to M.

Following the general philosophy (cp. 8) there should exist infinite dimensional analogues of (11.1) and (11.4), e.g. with L and M lying in the infinite dimensional algebra $L_\Lambda \otimes \mathrm{gl}(2, \mathbb{C})$. Specifically [6]:

$$L = \lambda^2 \begin{pmatrix} 1 & 0 \\ 0 & -1 \end{pmatrix} + \lambda \begin{pmatrix} 0 & -\phi^* \\ \phi & 0 \end{pmatrix} - \frac{1}{2} \begin{pmatrix} -\phi^* \circ \phi & \Pi^* \\ \Pi & \phi \circ \phi^* \end{pmatrix} ,$$

$$M = -\lambda \begin{pmatrix} 1 & 0 \\ 0 & -1 \end{pmatrix} - \begin{pmatrix} 0 & -\phi^* \\ \phi & 0 \end{pmatrix} , \tag{11.7}$$

$$\dot{L} = [L, M] := L * M - M * L ,$$

where ϕ and Π (taking the role of Q and P, respectively) are elements of the associative algebra A_Λ (underlying L_Λ), '$\circ$' denotes the multiplication in A_Λ,

$$T_\mathbf{m} \circ T_\mathbf{n} = \frac{i}{\lambda} \omega^{-\frac{1}{2}\mathbf{m} \times \mathbf{n}} T_{\mathbf{m}+\mathbf{n}} , \tag{11.8}$$

and '$*$' the composition of '$\circ$' with normal matrix multiplication. Using the representation (10.22) for the generators of L_Λ,

$$T_\mathbf{m} = \frac{i}{\lambda} \omega^{\frac{1}{2} m_1 m_2} e^{i m_1 \theta} e^{m_2 \lambda \partial_\theta} ,$$

$$\omega = e^{i\lambda} , \quad \lambda = 4\pi\Lambda , \tag{11.9}$$

the Hamiltonian, which abstractly is defined as

$$H = \frac{1}{2} \operatorname{Tr} \left(\Pi^+ \circ \Pi + \phi^+ \circ \phi \circ \phi^+ \circ \phi \right) \tag{11.10}$$

— Tr being an invariant trace in A_Λ — becomes

$$H = \frac{1}{2} \int_0^{2\pi} \frac{d\theta}{2\pi} \left(p_m(\theta) p_m^*(\theta) \right.$$

$$\left. + \sum_{m-n+k-l=0} q_m(\theta) q_n^* (\theta + \lambda(m-n)) \, q_k (\theta + \lambda(m-n)) \, q_l^*(\theta) \right) \tag{11.11}$$

upon

$$\phi = \sum_m q_m(\theta, t) T^m \quad , \quad \Pi = \sum p_m T^m \quad ; \quad T = e^{\lambda \partial_\theta} \quad . \tag{11.12}$$

The equations of motion read (omitting t in $q_m(\theta, t)$)

$$\ddot{q}_m(\theta) = -2 \sum_{n,k} \left(q_{m+n-k}(\theta) q_n^*(\theta + \lambda(m-k)) q_k(\theta + \lambda(m-k)) \right) \quad . \tag{11.13}$$

Unfortunatly, (11.7) has not (yet) been generalized to yield

$$H = \frac{1}{2} \text{Tr} \left(\Pi^+ \circ \Pi + (\phi^+ \circ \phi - \phi \circ \phi^+)^2 \right) \quad , \tag{11.14}$$

which in the limit $\lambda \to 0$ would give

$$H = \frac{1}{2} \int \frac{d\varphi_1 d\varphi_2}{(2\pi)^2} \left(\Pi^* \cdot \Pi - \{\phi^*, \phi\}^2 \right) \quad , \tag{11.15}$$

with $\{ , \}$ denoting the Poisson–bracket for functions on the torus. Up to an additional constraint (that projects onto some invariant subsystem) (11.15) describes the dynamics of a relativistically invariant bosonic membrane moving in four space–time dimensions. Written in real components,

$$\phi = \frac{1}{\sqrt{2}} (q_1(\varphi) + i q_2(\varphi)) \quad , \quad \Pi = \frac{1}{\sqrt{2}} (p_1(\varphi) + i p_2(\varphi)) \quad , \tag{11.16}$$

Eq. (11.15) reads

$$H = \frac{1}{2} \int \frac{d\varphi_1 d\varphi_2}{(2\pi)^2} \left(p_1^2 + p_2^2 + \{q_1, q_2\}^2 \right) \quad . \tag{11.17}$$

If one of the two fields q_i $(i = 1, 2)$ is made non–dynamical, one gets an integrable system, namely

$$H = \frac{1}{2} \int \frac{d\varphi_1 d\varphi_2}{(2\pi)^2} \left(p^2 + \{q, \omega\}^2 \right) \quad , \tag{11.18}$$

with $\omega = \omega(\varphi_1, \varphi_2)$ (formerly q_2) playing the role of a time–independent external field.

Somewhat surprisingly (though logically more or less following from (10.10)) the integrability of (11.18) does not rely on the specific (effectively quadratic) interaction; rather one can take

$$\begin{aligned} L &= p(\varphi_1, \varphi_2) - f'(\varphi_3) F(\{q, \omega\}) \quad , \\ M &= -f(\varphi_3) F(\{q, \omega\}) \end{aligned} \tag{11.19}$$

as a Lax–pair for

$$H = \frac{1}{2} \int \frac{d\varphi_1 d\varphi_2}{(2\pi)^2} \left(p^2(\varphi_1, \varphi_2) + \lambda F^2(\{q, \omega\}) \right) \tag{11.20}$$

for any F (and f) satisfying

$$FF'' = \alpha F'^2$$
$$f'^2 - \alpha f f'' = \lambda(\alpha + 1) = \text{const}.$$

(11.21)

Note that altough locally one could choose coordinates such that (11.20) is merely a non–dynamical addition of $1 + 1$ dimensional models (10.10), this is not true globally!

In any case, one can easily show that the equations of motion following from (11.20) are equivalent to

$$\dot{L} = \epsilon_{abc} \partial_a L \partial_b \omega \partial_c M \quad ,$$

(11.22)

which in particular implies that all

$$Q_{kl} := \int d\varphi_1 d\varphi_2 d\varphi_3 L^k \omega^l$$

are conserved charges.

References

[1]   A. Fordy, S. Wojciechowsky, I. Marshall; Phys. Lett. A 113 (86) 395.
[2]   R. Garnier; Rend. Circ. Math. Palermo 43 #4 (1919) 155.
[3]   A. Reyman; vii Zap. Nauch. Semin. LOMI 155 (86) 187.
[4]   I. Marshall; Phys. Lett. A 127 (1988) 19.
[5]   S. Helgason; *Differential Geometry and Symmetric Spaces*, Academic Press 1962.
[6]   J. Hoppe; Phys. Lett. B 250 (1990) 44.

12 Differential Lax Operators

Instead of considering Lax pairs of infinite dimensional matrices (implying specific basis in vector spaces of countable dimension) one often looks at Lax equations

$$\dot{L} = [L, M] \quad , \tag{12.1}$$

where L and M are differential operators of order l and m, respectively,

$$L = D^l + \sum_{i=0}^{l-2} u_i(x,t)D^i$$

$$M = \sum_{i=0}^{m} v_i(x,t)D^i \quad , \quad D = \frac{\partial}{\partial x} \quad . \tag{12.2}$$

As the r.h.s. of (12.1) will generically be of order $m + l - 1$, while $\dot{L}$ is at most of order $l - 1$, the functions v_i have to be choosen quite particularly (as polynomials in the u_i and their derivatives). As it turns out [1], M should be chosen as the positive part of any integer power of $L^{1/l}$, which is well defined by going to the algebra $\mathcal{P}$ of 'pseudodifferential operators',

$$P = \sum_{i=0}^{p} P_i D^i + \sum_{i=1}^{\infty} p_i D^{-i} = P_+ + P_- \quad . \tag{12.3}$$

Obviously, one needs to define D^{-1}; requiring $D^{-1} \cdot D = D \cdot D^{-1} = \mathbb{1}$ and associativity of $\mathcal{P}$, one finds

$$D^{-m} f = fD^{-m} + \sum_{i=1}^{\infty} (-1)^i \binom{m+i-1}{i} f^{(i)} D^{-i-m} \quad , \quad m > 0 \quad . \tag{12.4a}$$

One may obtain (12.4a) by induction in m, starting with $D^{-1} \cdot f = f \cdot D^{-1} + X[f]$, i.e. $X[f] = -D^{-1} f' D^{-1}$, which can be solved by iteration. Also note that when writing the binomial coefficient in (12.4a) as $\gamma_{mi} = (m+i-1)(m+i-2)\cdots(m)/i!$, the equation also holds for $m = -n < 0$, yielding the usual formula

$$[D^n, f] = \sum_{i=0}^{n-1} \binom{n}{i} f^{n-i} D^i \quad , \quad n \geq 0 \quad . \tag{12.4b}$$

Combining (12.4a/4b), one has

$$[D^n, f] = \sum_{i=1}^{\infty} \frac{n(n-1)\cdots(n-i+1)}{i!} f^{(i)} D^{n-i} \quad , \quad n \in \mathbb{Z} \tag{12.4}$$

for *all* n.

The fact, that L (as in (12.2)) has a unique l–th root of the form

$$L^{1/l} = D + \sum_{i=1}^{\infty} l_i(x)D^{-i} \quad , \tag{12.5}$$

was already observed by Schur [2], and can be proven by simply writing out $\left(L^{1/l}\right)^l = L$ and equating coefficients of D^j. Also it is not difficult to see that

$$M = \left(L^{m/l}\right)_+ \tag{12.6}$$

(with '+' denoting that part of a pseudodifferential operator that does not contain negative powers of D) will be consistent with (12.1): as $L^{1/l}$ commutes with $L = \left(L^{1/l}\right)^l$, hence L with $L^{m/l}$, one knows that $\left[L, \left(L^{m/l}\right)_+\right] = \left[\left(L^{m/l}\right)_-, L\right]$, which is of the order $l - 2$, as $\left(L^{m/l}\right)_-$ by definition contains only negative powers of D. Note that this explains also why the coefficient functions of D^{l-1} in L had been assumed to be zero; the above argument would have implied the non–dynamicality of that function, anyway.

Having thus found a consistent way to choose M (given L) in a non–trivial way (there is *one* for every positive integer that is not a multiple of l) the next step is to deduce conservation laws from (12.1). For this one would 'just' need an invariant trace (recall that (12.1) implies $d/dt(\text{Tr } L^n) = \text{Tr }[L^n, M] = 0$, if $\text{Tr } AB = \text{Tr } BA$). In the case of differential operators, however, no such trace operation seems to exist. Instead, one notes that (12.1) implies (as an identity in $\mathcal{P}$)

$$\frac{d}{dt}\left(L^{r/l}\right) = \left[L^{r/l}, M\right] \quad , \quad r \in \mathbb{N} \tag{12.7}$$

and that for *pseudodifferential* operators there does exist a non–trivial invariant trace, namely (cp. (12.3))

$$\text{Tr } P = \int p_1(x)dx \quad . \tag{12.8}$$

p_1 is called the 'residue' of P, and denoted by $\text{Res } P$. Using (12.4), it is not difficult to see that

$$\text{Tr}[P, Q] = 0 \quad \forall_{P,Q \in \mathcal{P}} \quad . \tag{12.9}$$

$$F = f(x)D^m \quad , \quad G = g(x)D^n \quad m, n \in \mathbb{Z}$$

$$F \cdot G = f D^m g D^n$$

$$= f \left(\sum_{i=0}^{\infty} \gamma_{m_i} g^{(i)} D^{m-i} \right) D^n$$

$$\text{with } \gamma_{m_i} = m(m-1)\cdots(m-i+1)/i! \text{ (and } \gamma_{m_0} := 1)$$

$$G \cdot F = g D^n f D^m$$

$$= g \left(\sum_{i=0}^{\infty} \gamma_{n_i} f^{(i)} D^{n-i} \right) D^m$$

$$\Rightarrow [F, G] = \sum_{i=1}^{\infty} \left(\gamma_{m_i} f g^{(i)} - \gamma_{n_i} f^{(i)} g \right) D^{m+n-i}$$

$$= \sum_{j=1-(m+n)}^{\infty} h_j D^{-j} \quad ; \quad h_1 =: \mathrm{Res}[F, G]$$

$$\mathrm{Res}[F, G] = 0 \quad \text{if } (m+n) < 0, \text{ or if } m, n \geq 0$$

$$\text{otherwise: } = \left(\gamma_{m,m+n+1} f g^{(m+n+1)} - \gamma_{n,m+n+1} f^{(m+n+1)} g \right)$$

$$= \gamma_{m,m+n+1} \left(f g^{m+n+1} - \left(\underbrace{\frac{n(n-1)\cdots(-m)}{m(m-1)\cdots(-n)}}_{=(-1)^{m+n+1}} f^{m+n+1} g \right) \right)$$

$$= \gamma_{m,m+n+1} \underbrace{\left(\sum_{l=0}^{m+n} (-1)^l f^{(l)} g^{(m+n-l)} \right)'}_{\parallel} =: (k(x))'$$

$$\left(f g^{(m+n)} - f' g^{(m+n-1)} + \ldots + (-1)^{m+n} f^{(m+n)} g \right)'$$

$$\Rightarrow \int dx \, \mathrm{Res}[F, G] = 0 \quad \forall_{m,n \in \mathbb{Z}} \quad .$$

This means that

$$Q_r = \int \mathrm{Res}(L^{\frac{r}{t}}) \quad , \, r \in \mathbb{N} \tag{12.10}$$

will be conserved,

$$\frac{d}{dt} Q_r = \int \mathrm{Res} \left[L^{\frac{r}{t}}, M \right] = 0 \quad , \tag{12.11}$$

irrespective of what power $m \in \mathbb{N}$ we choose to define M (cp.(12.6)). As the variable t only enters via (12.1), one may therefore associate *different* 'time'–variables to each choice of M,

$$\frac{d}{dt_m} L^{\frac{r}{t}} = \left[L^{\frac{r}{t}}, (L^{\frac{m}{t}})_+ \right] \quad , \quad m, r \in \mathbb{N} \quad . \tag{12.12}$$

Instead of considering each m seperatly (i.e. $u_i = u_i(x, t_m)$) one considers them all at once, i.e. u_i $(i = 0, \ldots, l-2)$ as functions of infinitely many 'time' variables t_m,

$$u_i = u_i(x = -t_1, t_2, t_3, \ldots) \quad . \tag{12.13}$$

For this to be possible, it is necessary to prove that (12.12) is consistent with

$$\frac{d^2}{dt_m dt_n} = \frac{d^2}{dt_n dt_m} \quad , \tag{12.14}$$

when acting on $L^{\frac{r}{l}}$.

Letting $N = \left(L^{\frac{n}{l}}\right)$, $K = \left(L^{\frac{k}{l}}\right)$, $\partial_n P = [P, N]$, $\partial_k P = [P, K]$, one finds (using only the Jacobi–identity) that

$$\left(\partial_{nk}^2 - \partial_{kn}^2\right) P = \left[P, [N_+, K_+] + [K, N_+]_+ - [N, K_+]_+\right] \quad . \tag{12.15}$$

$[K, N] = 0$, on the other hand, implies that $[K, N_+]_+ = -[K, N_-]_+ = -[K_+, N_-]_+$, which shows that the r.h.s. of (12.15) is identically zero for any $P \in \mathcal{P}$.

The infinite set of equations (or rather the infinitely many physical systems, or nonlinear evolution equations, defined by)

$$\partial_m L = \left[L, \left(L^{\frac{m}{l}}\right)_+\right] \quad m \in \mathbb{N} \tag{12.16}$$

are called the 'L–hierarchy'.

To each L (determining the hierarchy) and $m \in \mathbb{N}$ (denoting a particular member in that hierarchy) there exist infinitely many conserved quantities, Q_r (independent of m),

$$\partial_m Q_r = 0 \quad m, r \in \mathbb{N} \quad . \tag{12.17}$$

Let us look at the simplest example ($l = 2$),

$$L = D^2 + u \quad ; \tag{12.18}$$

as the first non–trivial member in this hierarchy,

$$\partial_3 L = \left[L, \left(L^{\frac{3}{2}}\right)_+\right] \quad , \tag{12.19}$$

is equivalent to the Kortweg de Vries (KdV) equation,

$$\dot{u} = -\frac{1}{4}(u''' + 6uu') \quad . \tag{12.20}$$

Eq. (12.16), with L as in (12.18), is called the KdV–hierarchy; it is the original, and also most famous example.

Let us calculate the first few conserved quantities for this case. Obviously, one needs $L^{\frac{1}{2}} = D + \sum_{i=1}^{\infty} l_i D^{-i}$ (cp. (12.5)). Comparing coefficients of $L^{\frac{1}{2}} \cdot L^{\frac{1}{2}} = D^2 + u$, one finds ($l_0(x) = 0$)

$$l_{n+1}(x) = -\frac{1}{2}\left\{l'_n(x) + \sum_{i+j=n} l_i l_j + \sum_{i+j+k=n}\sum_{k=1}^{n-1}(-1)^k l_i l_j^{(k)}\binom{i+k-1}{k}\right\} ,$$

$$(12.21)$$

$$L^{\frac{1}{2}} = D + \frac{1}{2}uD^{-1} - \frac{1}{4}u'D^{-2} + \frac{1}{8}\left(u'' - u^2\right)D^{-3} + \frac{3}{8}\left(uu' - \frac{1}{2}u'''\right)D^{-4} + \cdots .$$

$$(12.22)$$

Taking the residue of $L^{\frac{1}{2}}$, as well as of

$$L^{\frac{3}{2}} = L \cdot L^{\frac{1}{2}} = D^3 + \frac{3}{2}uD + \frac{3}{4}u' + \left(\frac{1}{8}u'' + \frac{3}{8}u^2\right)D^{-1} + \cdots , \qquad (12.23)$$

and $L^{\frac{5}{2}} = L \cdot L \cdot L^{\frac{1}{2}}$, yields

$$Q_1 = \frac{1}{2}\int u ,$$

$$Q_3 = \frac{3}{8}\int u^2 ,$$

$$(12.24)$$

$$Q_5 = \frac{5}{16}\int\left(u^3 - \frac{1}{2}u'^2\right) ,$$

$$Q_2 = Q_4 = \cdots = 0 .$$

To verify that $\partial_3 Q_m = 0$ (e.g.), it suffices to directly use the relevant equation(s) of motion, (e.g.)(12.20).

One should mention that there exist more sophisticated methods of calculating the conserved charges Q_r; starting with (12.5) and writing

$$D = L^{\frac{1}{r}} + \sum_1^\infty f_i L^{-\frac{i}{r}} \qquad (12.25)$$

one can show that

$$-r\int f_r dx = Q_r = \int \text{Res } L^{\frac{r}{r}} . \qquad (12.26)$$

To obtain the f_i, one uses formal eigenfunctions ψ_λ,

$$L^{\frac{1}{r}}\psi_\lambda(x) = \lambda\psi_\lambda(x) \qquad (12.27)$$

which, according to (12.25), satisfy

$$\psi'_\lambda(x) = \left(\lambda + \sum_1^\infty f_i\lambda^{-i}\right)\psi_\lambda . \qquad (12.28)$$

Let us calculate the first few f_i from (12.28), in the example of the KdV Lax–operator (12.18). Observing that $(D^2 + u)\psi_\lambda = \lambda^2\psi_\lambda$ implies

$$\phi' + \phi^2 = \lambda^2 - u \quad ; \quad \phi_\lambda := \frac{\psi'_\lambda}{\psi_\lambda} \quad , \tag{12.29}$$

and assuming an expansion of ϕ in negative powers of λ, according to (12.28), one derives the recursion formula

$$f_{n+1} = -\frac{1}{2} \left(f'_n + \sum_{i+j=n} f_i f_j \right) \tag{12.30}$$

$$f_1 = -\frac{1}{2} u \quad .$$

So

$$f_2 = \frac{1}{4} u' \,, \qquad\qquad f_3 = -\frac{1}{8}(u^2 + u'')\,,$$

$$f_4 = \frac{1}{16}(u''' + 4uu')\,, \; f_5 = -\frac{1}{2}\left(f'_4 + \frac{u'^2}{16} + \frac{1}{8}\left(u^3 + uu''\right)\right)\,, \tag{12.31}$$

verifying (12.26/27) for $r < 6$.

One more remark on hierarchies: if the r.h.s. of (12.5) is not required to be the l–th root of some ordinary differential operator, i.e. for a *general* pseudo–differential operator P of the form

$$P = D + \sum_1^\infty p_i D^{-i} \tag{12.32}$$

one may still consider

$$\partial_m P = [P, (P^m)_+] \quad m \in \mathbb{N} \tag{12.33}$$

(obviously one has to identify t_1 with $-x$). Equation (12.33) is equivalent to the system of equations

$$\partial_m P_n - \partial_n P_m + [P_m, P_n] = 0 \quad ; P_m = (P^m)_+ \quad . \tag{12.34}$$

As one of these, $(m = 2, n = 3)$, yields the Kadomtsev–Petviashvili (KP) equation,

$$3u_{yy} + (4u_t + u_{xxx} + 6uu_x)_x = 0 \tag{12.35}$$

for $u = \mathrm{Res}\, P$ (and, up to signs, $y = t_2$ and $t = t_3$), (12.33) is called the KP–hierarchy. It contains all hierarchies of type (12.16), i.e. there exist *special* solutions of (12.32) for which the p_i are such that a given positive power of P is an ordinary differential operator.

Notes and References

For more details, proofs and much further information on the KdV–hierarchy, and related topics, see [1]. A short review of the KP–hierarchy is contained in [3], which otherwise presents the Toda–lattice hierarchy (making use of $gl_J(\infty)$ instead of the algebra of pseudo–differential operators).

[1]   V.G. Drinfeld, V.V. Sokolov; Journal of Sov. Math. 30 (1985) 1975.
[2]   I. Schur; *Über vertauschbare lineare Differentialausdrücke* (1904), Sitzungsbericht
        d. Berliner Math. Gesellschaft (1905) 2–8.
[3]   K. Ueno, K. Takasaki; Adv. Stud. in Pure Math. 4 (1984) 1.

13 First Order Differential Matrix Lax Operators and Drinfeld–Sokolov Reduction

Up to now, the Lax operator L was a higher order ($l > 1$) differential operator, with scalar coefficient functions u_i ($i = 0, \ldots, l - 2$). For a suitable formulation of consistent Lax equations, and the formulation of the corresponding conserved quantities, pseudo–differential operators were needed, in particular their decomposition into positive (=ordinary differential) and negative part, and the grading given by the power of ∂_x.

Analogous considerations for first order (in ∂_x) matrix Lax operators with spectral parameter λ (Laurent–series in λ with matrix-valued coefficients, and a grading by the power of λ, playing the role of pseudo–differential operators and ∂_x) yield the following result [1]:

$$\dot{L} = [L, M] \tag{13.1}$$

$$
\begin{aligned}
L &= \mathbb{1} \cdot \partial_x + Q - \lambda C \\
Q &= (q_{rs})_{r,s=1,2,\ldots,N} = Q(x,t) \\
C &= \operatorname{diag}(c_1, c_2, \ldots, c_N) \\
M &= \sum_{i=0}^{m} M_i \lambda^i
\end{aligned}
\tag{13.2}
$$

presents a consistent set of nonlinear partial differential equations for the $q_{rs}(x,t)$, if M is chosen as follows:

— choose m+1 constant diagonal matrices $D_0, D_1, \ldots, D_m$
  $D = \sum_0^m D_i \lambda^i$
— consider $X = T^{-1}DT$, where $T = \mathbb{1} + \sum_{i=1}^{\infty} T_i \lambda^{-i}$ is choosen such that

$$
\begin{aligned}
TL &= (\partial_x - \lambda C + H)T \\
H &= \sum_0^{\infty} H_i \lambda^{-i} = \operatorname{diag}.
\end{aligned}
\tag{13.3}
$$

— Set

$$M = \left(T^{-1}DT\right)_+ , \tag{13.4}$$

where $_+$ denotes the projection onto the part containing non–negative powers of λ.

T and H are recursively (and uniquely, if $(T_i)_{jj} = 0$) defined by (13.3),

$$H_n + [T_{n+1}, C] = -T_n' + T_n Q - \sum_{j=1}^n H_{n-j} T_j \quad n \geq 0 \quad . \tag{13.5}$$

Furthermore,

$$\frac{d}{dt} \int H(x, \lambda) dx = 0 \quad , \tag{13.6}$$

so that $\int H(x, \lambda) dx$ is a generating (matrix) function (of λ) for an infinite set of time–independent quantities. This is implied by the diagonal part of

$$0 = \left[T \left(\partial_t + M \right) T^{-1}, \partial_x - \lambda C + H \right]$$
$$= \left[\partial_t + \tilde{M}, \partial_x + \tilde{H} \right] \tag{13.7}$$
$$= \partial_t \tilde{H} - \partial_x \tilde{M} + \left[\tilde{M}, \tilde{H} \right]$$

Example: N–wave equation

Choose $m = 1$, $D_0 = 0$, i.e.

$$D = \lambda D_1 = \lambda \cdot \mathrm{diag}(d_1, \ldots, d_N) \quad . \tag{13.8}$$

So

$$M = (T^{-1} D T)_+$$
$$= \left(\left(\mathbb{1} - \frac{T_1}{\lambda} + \ldots \right) D_1 \lambda \left(\mathbb{1} + \frac{T_1}{\lambda} + \ldots \right) \right)_+ = \lambda D_1 + [D_1, T_1] \quad . \tag{13.9}$$

As (from (13.5), i.e. $H_0 + [T_1, C] = Q$)

$$(T_1)_{r \neq s} = -\frac{q_{rs}}{C_r - C_s} \tag{13.10}$$

$$M_{rs} = \lambda \delta_{rs} d_r - a_{rs} q_{rs}$$
$$a_{r \neq s} = \frac{d_r - d_s}{C_r - C_s} \quad , \quad a_{rr} \equiv 0 \quad , \tag{13.11}$$

and (inserting (13.11) into (13.1), checking explicitly that λ–dependent terms cancel)

$$\dot{q}_{rs} = -a_{rs} q_{rs}' + \sum_t (a_{rt} - a_{ts}) q_{rt} q_{ts} \quad . \tag{13.12}$$

Restricting (13.12) to antisymmetric, x–independent Q and $N = 3$ would give the Euler–equations of motion for a free top.

Note that if one is not interested in the explicit form of the conserved quantities, the prescription (13.2-5) could have been replaced by a recursive ad hoc determination of $M = \sum_i M_i \lambda^i$ such that in $[L, M]$ terms with positive powers of λ cancel.

Drinfeld Sokolov Reduction

Let us now show a relation between first order differential matrix operators L and higher order differential scalar L's. For this, following [1], one starts with a form slightly different from (13.1),

$$L = \partial_x + Q + \Lambda \quad , \qquad Q = \begin{pmatrix} q_{11} & & & q_{i<j} \\ & q_{22} & & \\ & & \ddots & \\ 0 & & & q_{NN} \end{pmatrix}$$

$$\Lambda = \begin{pmatrix} 0 & & & 0 \\ 1 & \ddots & & \\ & \ddots & \ddots & \\ 0 & & 1 & 0 \end{pmatrix} + \lambda \begin{pmatrix} 0 & \cdots & 0 & 1 \\ 0 & \cdots & 0 & 0 \\ & & \vdots & \\ 0 & \cdots & & 0 \end{pmatrix} = \Lambda_- + \lambda \cdot I \quad ,$$

(13.13)

i.e. Q upper triangular (u.t.) and λ entering in a very specific 'unsymmetrical' way, noting that similarity transformations $L \to R^{-1}LR$ with special upper triangular (s.u.t.) matrices

$$R = \begin{pmatrix} 1 & & r_{i<j} \\ & \ddots & \\ 0 & & 1 \end{pmatrix}$$

(13.14)

leave invariant the form of L:

$$R^{-1}LR = \partial_x + \tilde{Q} + \Lambda \quad , $$

(13.15)

where $\tilde{Q}$ is again upper triangular. This is easily seen by noting $R^{-1} =$s.u.t., $R^{-1}IR = I$, $R^{-1}\Lambda_- R = \Lambda_- +$u.t., $R^{-1}\partial_x R$ and $R^{-1}QR$ both u.t.. As R contains $N(N-1)/2$ free functions, one may try, using the 'gauge-transformations' (13.15), to put all but N of the $N(N+1)/2$ functions $q_{i\le j}$ equal to zero. Indeed, there exists an R (s.u.t.) such that in (13.15)

$$\tilde{Q} = \bar{Q} = \begin{pmatrix} 0 & \cdots & 0 & v_1 \\ 0 & \cdots & 0 & v_2 \\ \vdots & & \vdots & \\ 0 & \cdots & 0 & v_N \end{pmatrix} .$$

(13.16)

This is most easily proven by writing in

$$(\partial_x + Q + \Lambda)R = R(\partial_x + \bar{Q} + \Lambda)$$

(13.17)

all occuring matrices as sums of their (off)diagonal 'slices',

$$Q = \sum_0^{N-1} Q_i \quad , \quad R = \mathbb{1} + \sum_1^{N-1} R_i \quad , \quad \bar{Q} = \sum_0^{N-1} \bar{Q}_i \quad ,$$

(13.18)

with $(M_i)_{jk} = M_{jk}\delta_{i+j,k}$ ($M = Q, R, \bar{Q}, \ldots$), and then recursively considering the n-th off-diagonal of (13.17), $n = 0, 1, \ldots, N-1$,

$$[R_{n+1}, \Lambda_-] + \bar{Q}_n = R'_n + Q_n + \sum_1^n Q_{n-j} R_j - \sum_1^n R_j \bar{Q}_{n-j} \quad . \tag{13.19}$$

As an example, the first step ($n = 0$) yields

$$r_{i,i+1} = \sum_{j=1}^i q_{jj} \quad (i \leq N-1) \quad , \quad v_n = \bar{q}_{NN} = \operatorname{Tr} Q \quad . \tag{13.20}$$

The next step ($n = 1$) results in $r_{i,i+2}$ ($i = 1, \ldots, N-2$) containing first derivatives (of q_{jj}'s), quadratic terms (in the q_{jj}'s) as well as terms linear in the $q_{j,j+1}; \ldots$ Obviously, all v_j ($1 \leq j \leq N$) will be polynomials in the $q_{i \leq j}$ and their derivatives. They will be gauge–invariant(!) in the sense that any $L' = SLS^{-1}$ yields the same v_i (this follows from the fact that the solution of (13.19) is *unique*, i.e. $L_1 = R\bar{L}_1 R^{-1}$, $L_2 = S\bar{L}_2 S^{-1} \overset{!}{=} TL_1 T^{-1} = (TR)\bar{L}_1(TR)^{-1}$ implies $\bar{L}_1 = \bar{L}_2$ and $S = TR$). Moreover, the v_i ($i = 1, \ldots, N$) generate a ring (J) of *all* gauge–invariant differential polynomials (in the matrix elements of Q).

Up to now, time has not entered. When making Q dynamical by prescribing some evolution equation $\frac{d}{dt} Q = F[Q]$, involving L, it would be nice to have *gauge–invariant* equations, meaning that if one calculates the time–derivative of $v_i = v_i[Q]$, $i = 1, \ldots, N$, by using how Q changes in time, the result should lie again in J. Such gauge–invariant equations *do* result from

$$\dot{L} = [L, M] \tag{13.21}$$

if M is chosen as follows: consider constant matrices

$$\tilde{C} = \sum_{-\infty}^m \tilde{c}_i \Lambda^i \quad , \quad \tilde{c}_i \in \mathbb{C} \quad . \tag{13.22}$$

Transform $\tilde{C}$ according to the similarity transformation that brings L into the form

$$TLT^{-1} = \partial_x + \Lambda + \sum_{i=0}^\infty f_i(x) \Lambda^{-i} = L_0$$

$$T = \sum_0^\infty T_i \lambda^{-i} \quad ; \quad f_i(x) \in \mathbb{C} \quad , \quad T_0(x) \text{ s.u.t.} \quad . \tag{13.23}$$

As $\tilde{C}$ commutes with L_0, hence $T^{-1}\tilde{C}T$ with L,

$$M = (T^{-1}\tilde{C}T)^+ \quad , \tag{13.24}$$

will have the property that $[L, M]$ will be upper triangular and not depend on λ. ($(X)^+$ denotes $\sum_{i>0} X_i \Lambda^i$, if $X = \sum_{i=-\infty}^n X_i \Lambda^i$, X_i diagonal.) Moreover, the resulting equations of motion are gauge–invariant, with equations for $v_1, \ldots, v_N$ that are identical with those resulting from

$$\dot{L} = [L, M]$$

$$L = \partial^N + \sum_{i=0}^{N-1} v_{i+1} \partial^i \tag{13.25}$$

$$M = (L^{m/N})_+ \quad .$$

A *different* reduction of the first order differential matrix operator $L = \partial_x + Q + \Lambda$ (cp. (13.13)),

$$R^{-1} L R = \partial_x + \hat{Q} + \Lambda \quad , \quad \hat{Q} = \text{diag}(q_1, \ldots, q_N) \tag{13.26}$$

can be shown to be equivalent to the N-th order scalar Lax–operator

$$L^{(N)} = (\partial - q_N)(\partial - q_{N-1}) \cdots (\partial - q_1) \quad . \tag{13.27}$$

This transition is called 'Miura–mapping' corresponding to the person who first considered such a relation, for the simplest example —

$$L = \partial_x + \begin{pmatrix} q & \lambda \\ 1 & -q \end{pmatrix} \quad , \quad L^{(2)} = (\partial_x + q)(\partial_x - q) = \partial_x^2 + u \quad , \tag{13.28}$$

$$u = -q' + q^2 \quad .$$

Notes and References

Zero curvature conditions of type (13.1), i.e. on finite dimensional matrix algebras, were formulated in [2], and used in the context of integrating non–linear partial differential equations by the method of inverse scattering. As indicated by its name, the best reference for the Drinfeld–Sokolov reduction is [1], especially with respect to Kac–Moody generalizations. Concerning Miura-maps see also [3,4] (and references therein) as well as (for a very recent context) [5].

[1]   V.G. Drinfeld, V.V. Sokolov; Journal of Sov. Math. 30 (1985) 1975.
[2]   V.E. Zakharov, S.V. Manakov; JETP 42 (1976) 842.
[3]   F. Calogero, A. Degasperis; *Solitons and the Spectral Transform I*, North Holland 1982.
[4]   A.C. Newell; *Solitons in Mathematics and Physics* CBMS–NSF Ref. Conf. Ser. in Appl. Math. 48 SIAM 1985
[5]   E. Kawai; Phys. Lett. B 271 (1991) 347.

14 Zero Curvature Conditions on W_∞, Trigonometrical and Universal Enveloping Algebras

The use of first order differential operators

$$L = 1 \cdot \partial_x + Q - \lambda C \qquad (14.1)$$

in the construction of non–linear evolution equations posessing an infinite number of conserved quantities involved a finite number of x and t dependent fields $q_{rs}(x,t)$ $(r,s = 1 \ldots N)$, that made up Q. The constant matrix C was chosen diagonal, and the construction of the constants of motion relied on the possibility to 'diagonalize' L by means of a similarity transformation R, which required the solution of certain recursion relations, (13.5).

As is not difficult to see, one may as well consider an infinite number of dynamical fields, by letting Q (and C), instead of $N \times N$ matrices, be elements of some infinite dimensional, non–commutative associative algebra $\mathcal{A}$ (over $\mathbb{C}$). In view of (13.5), and with $\mathcal{H} \subset \mathcal{A}$ being some (maximal) abelian subalgebra, $\mathcal{A}$ should have the property that for suitably chosen (fixed) $C \in \mathcal{H}$

$$\begin{array}{cc} & X \in \mathcal{A} \backslash \mathcal{H} \\ H + [C,X] = Y & Y \in \mathcal{A} \\ & C, H \in \mathcal{H} \end{array} \qquad (14.2)$$

is (uniquely) solvable for H and X, for any $Y \in \mathcal{A}$ (the uniqueness would corespond to a direct sum decomposition $\mathcal{A} = \mathcal{H} \oplus [A,C]$, the centralizer of C coinciding with $\mathcal{H}$).

Let us consider three (classes of) examples, A^λ, $(\lambda \in \mathbb{C})$, $A_{\tilde{A}}$ ($\tilde{A}$ a complex, antisymmetric $N \times N$ matrix) and W_∞.

First Example: A^λ is defined by dividing out of U, the enveloping algebra of $sl(2,\mathbb{C})$, the ideal generated by

$$T_+ T_- - T_3^2 + T_3 - \lambda 1 \quad , \qquad (14.3)$$

where T_+, T_- and T_3 are a basis of $sl(2,\mathbb{C})$, satisfying

$$[T_3, T_\pm] = \pm T_\pm \quad , \quad [T_+, T_-] = -2T_3 \quad . \qquad (14.4)$$

For $\lambda \neq \frac{1}{4}(1 - n^2)$, $n \in \mathbb{N}_0$, the A^λ are simple and non–isomorphic for different values of $\lambda \in \mathbb{C}$. As a basis of A^λ one may take

$$e_{mk} = T_\pm^m T_3^k \quad (m, k \in \mathbb{N}_0) \quad ,$$

rewriting monoms containing T_+ and T_- by using that $(14.3) = 0$ in A_λ; as C we may take $C = T_3$. It is easy to see that (14.4) implies

$$T_\pm^m T_3^k = \delta_{k,0} T_3^k \pm \frac{1 - \delta_{m,0}}{m} [T_3, T_\pm^m T_3^k] \quad . \tag{14.5}$$

Second Example: The second class of examples may (e.g.) be understood as coming from projective representations of the commutative group $\mathbf{Z}^d$, $d \geq 2$:

$$T(\mathbf{m}) \cdot T(\mathbf{n}) = e^{i\beta(\mathbf{m},\mathbf{n})} T(\mathbf{m} + \mathbf{n})$$

$$\beta(\mathbf{m}, \mathbf{n}) = 2\pi \mathbf{m}^t \tilde{\Lambda} \mathbf{n} \quad ; \quad \mathbf{m}, \mathbf{n} \in \mathbf{Z}^d \quad ; \tag{14.6}$$

$\tilde{\Lambda}$, an antisymmetric $d \times d$ matrix should, for our purpose, be chosen such that β never takes the value 0 $(\bmod\, 2\pi)$, unless $\mathbf{m} + \mathbf{n} = \mathbf{0}$. $A_{\tilde{\Lambda}}$ is then defined as the linear span over $\mathbb{C}$ of the $T(\mathbf{m}), \mathbf{m} \in \mathbf{Z}^d$, with the multiplication given by (14.6). As (14.6) implies

$$[T(\mathbf{m}), T(\mathbf{n})] = 2i \sin(2\pi \mathbf{m}^t \tilde{\Lambda} \mathbf{n}) \cdot T(\mathbf{n} + \mathbf{m}) \quad , \tag{14.7}$$

the centralizer of any $C = T(\mathbf{m})$ consists of linear combinations of $T(\mathbf{k} \parallel \mathbf{m})$, and any $T(\mathbf{n})$ may be written as

$$T(\mathbf{n}) = \begin{cases} T(\mathbf{n}) & \text{if } \mathbf{n} \parallel \mathbf{m} \\ \dfrac{1}{2i \sin 2\pi(\mathbf{m}^t \tilde{\Lambda} \mathbf{n})} [T(\mathbf{m}), T(\mathbf{n} - \mathbf{m})] & \text{if } \mathbf{n} \nparallel \mathbf{m} \quad , \end{cases} \tag{14.8}$$

which shows that the property (14.2) holds in $A_{\tilde{\Lambda}}$, with

$$\mathcal{H} = \mathcal{H}_{\mathbf{m}} = \left\{ \sum_{\mathbf{n} \parallel \mathbf{m}} h_{\mathbf{n}} T(\mathbf{n}) | h_{\mathbf{n}} \in \mathbb{C} \right\} \quad .$$

Realisations of $A_{\tilde{\Lambda}}$ and A_λ are given by

$$T(\mathbf{m}) = e^{\Sigma_{j=1}^d m_j B_j}$$

$$[B_j, B_k] = i4\pi \tilde{\Lambda}_{jk} \cdot \mathbb{1} \quad , \tag{14.9}$$

respectively

$$T_3 = \begin{pmatrix} b_1 & & 0 \\ & b_2 & \\ 0 & & \ddots \end{pmatrix} \quad ,$$

$$T_+ = \begin{pmatrix} 0 & & & 0 \\ a_1 & 0 & & \\ & a_2 & \ddots & \\ 0 & & \ddots & \end{pmatrix} \quad , \quad T_- = \begin{pmatrix} 0 & a_1 & & 0 \\ & 0 & a_2 & \\ & & \ddots & \ddots \\ 0 & & & \end{pmatrix} \quad , \tag{14.10}$$

$$a_n = \sqrt{n^2 + n(2k - 1)} \quad , \quad b_n = (n + k - 1)$$

$$\lambda = -k(k - 1) \leq \frac{1}{4} \quad .$$

Finally, $\mathbf{W}_\infty$ may conveniently defined (over $\mathbb{C}$) as a central extension of the Lie algebra $\mathcal{L}_0$ of differential operators (on S^1) with zero constant term and trigonometric polynomials as coefficient functions; as a basis of $\mathcal{L}_0$ (as well as the underlying associative algebra $\mathcal{A}_0$) one may take

$$T_{mk} = e^{im\theta}(-i\partial_\theta)^k \quad , \quad \begin{array}{c} m \in \mathbb{Z} \\ k = 1, 2, \ldots \\ \theta \in [0, 2\pi] \end{array} \tag{14.11}$$

satisfying

$$T_{mk} \cdot T_{nl} = \sum_{j=0}^{k} \binom{k}{j} n^j T_{m+n,k+l-j} \quad . \tag{14.12}$$

$\mathcal{L}_0$ has a unique non–trivial central extension, $\hat{\mathcal{L}}_0 \, (= W_\infty)$, given by

$$[\hat{T}_{mk}, \hat{T}_{nl}] = \sum_{j=1}^{r} f_{kl}^j(m, n)\hat{T}_{m+n,k+l-j}$$

$$+ c\frac{k!l!}{(k+l+1)!} m^l n^{k+1} \delta_{m+n,0} \tag{14.13}$$

$$f_{kl}^j(m, n) = \binom{k}{j} n^j - \binom{l}{j} m^j$$

$$\binom{x}{j} := 0 \text{ if } j > x \geq 0 \quad , \quad r = \max(k, l)$$

(Note the grading with respect to the first index). The T_{mk} (as in (14.11), and with $[\ ,\ ]$ being the natural commutator of the associative product (14.12)) satisfy (14.13), with $c = 0$.

Instead of using trigonometric polynomials and ∂_θ one often chooses to write the generators in terms of ∂_z and powers of $z = e^{i\theta}$. In any case, taking appropriate linear combinations of the T_{mk} (m fixed) one can define a basis T'_{mk} for which the central extension becomes 'diagonal', i.e. proportional to $\delta_{m+n,0}$ and $\delta_{k,l}$, and in which the structure constants coincide (up to normalization of the generators) with the ones originally found in [8] — where the realization of W_∞ in terms of differential operators was not used, rather the motivation (and interpretation) of generators $T_m^{(s)}$, $s = 2, 3, \ldots$, as the m–th Fourier mode of a field of conformal spins s.

In any case, to illustrate the type of dynamical equations which one may get by using these algebras, I will use the example $A_{\tilde{\Lambda}}$, $d = 2$, for which one has, slightly more explicit,

$$T_{\mathbf{m}} = \frac{i}{\lambda} e^{im_1\theta + m_2\lambda\frac{\partial}{\partial\theta}} = \frac{i}{\lambda}\omega^{\frac{1}{2}m_1 m_2} e^{im_1\theta} \left(e^{\lambda\partial_\theta}\right)^{m_2}$$

$$\omega = e^{i\lambda} \quad , \quad \tilde{\Lambda} = \frac{\lambda}{4\pi}\begin{pmatrix} 0 & -1 \\ 1 & 0 \end{pmatrix} \tag{14.14}$$

acting on smooth periodic functions of θ, or, acting as integral operators (on a somewhat larger space),

$$T_{\mathbf{m}} = \frac{i}{\lambda}\omega^{\frac{1}{2}m_1 m_2} e^{im_1 \theta} T^{m_2}$$

$$Tf(\theta) = f(\theta + \lambda)T \quad .$$

(14.14)

It is easy to check that

$$T_{\mathbf{m}} \cdot T_{\mathbf{n}} = \frac{i}{\lambda}\omega^{-\frac{1}{2}\mathbf{m} \times \mathbf{n}} T_{\mathbf{m}+\mathbf{n}} \quad ,$$

(14.15)

so $T_{\mathbf{m}} = \frac{i}{\lambda}T(\mathbf{m})$.

The most suggestive representation is probably

$$T_{\mathbf{m}} = \frac{i}{\lambda}e^{im\varphi}$$

(14.16)

in which $A_{\hat{\lambda}}$ is represented by smooth functions on the torus T^2 — thus Q by an ordinary function of four variables, $Q = f(x, t; \boldsymbol{\varphi})$, and with the non–commutative multiplication defined by

$$f(\boldsymbol{\varphi}) * g(\boldsymbol{\varphi}) = f \cdot g + \sum_{n=1}^{\infty} \frac{\left(\frac{i\lambda}{2}\right)^n}{n!} \epsilon_{r_1 s_1} \cdots \epsilon_{r_n s_n} \partial^n_{r_1 \ldots r_n} f \partial^n_{s_1 \ldots s_n} g \quad ,$$

(14.17)

or a corresponding integral formula that extends to discontinuous functions. $\mathcal{H}$ may in this case be represented by functions of *one* variable, φ_1. So let us write

$$L = \partial_x + f(x, t, \boldsymbol{\varphi}) + \mu\, c(\varphi_1)$$

(14.18)

and take the simplest possibility for M, $m = 1$,

$$M = g(x, t; \boldsymbol{\varphi}) + \mu\, b(\varphi_1) \quad ;$$

(14.19)

the $O(\mu)$ term is already chosen such that the $O(\mu^2)$ part of

$$\dot{L} = [L, M]$$

(14.20)

is automatically satisfied. Comparing terms linear in μ yields

$$[f, b]_* + [c, g]_* = 0 \quad ,$$

(14.21)

while the μ–independent part reads

$$\dot{f} = g' + f * g - g * f \quad .$$

(14.22)

Using

$$e^{im\varphi_2} * h(\varphi_1) = e^{im\varphi_2} h\left(\varphi_1 + \frac{m\lambda}{2}\right) = h(\varphi_1 + m\lambda) * e^{im\varphi}$$

$$f(\varphi_1) * g(\varphi_1) - g(\varphi_1) * f(\varphi_1) = 0 \quad ,$$

(14.23)

and writing

$$f(\varphi) = \sum_m f_m(\varphi_1) * e^{im\varphi_2} = \sum_m f_m\left(\varphi_1 - \frac{m\lambda}{2}\right) e^{im\varphi_2}$$

$$g(\varphi) = \sum_m g_m(\varphi_1) * e^{im\varphi_2} \quad,$$

(14.24)

(14.21) can easily be solved, yielding

$$g(\varphi) = \sum_m (a_m(\varphi_1) f_m(\varphi_2)) * e^{im\varphi_2}$$

$$a_m(\varphi_1) = \frac{b(\varphi_1) - b(\varphi_1 + m\lambda)}{c(\varphi_1) - c(\varphi_1 + m\lambda)} \quad.$$

(14.25)

Let us now look at the conserved charges. Just as in the finite–dimensional case, there exists (the x– and t–dependence will be suppressed)

$$r(\varphi) = 1 + \sum_1^\infty r_i(\varphi)\mu^{-i}$$

(14.26)

such that

$$r * L = (\partial_x + \mu c(\varphi_1) + h(\varphi_1)) * r$$

$$\text{and } h(\varphi_1) = \sum_0^\infty h_i(\varphi_1)\mu^{-i} \quad;$$

(14.27)

the r_i and h_i are recursively determined by

$$h_n(\varphi_1) + [c(\varphi_1), r_{n+1}(\varphi)]_*$$

$$= -r'_n + r_n * f - \sum_{j=1}^n h_{n-j} * r_j \quad.$$

(14.28)

As

$$\left[c(\varphi_1), \sum_m r_m^{(n+1)}(\varphi_1) * e^{im\varphi_2}\right]$$

$$= \sum_m \left(r_m^{(n+1)}(\varphi_1)(c(\varphi_1) - c(\varphi_1 + m\lambda))\right) * e^{im\varphi_2}$$

(14.29)

is zero when integrated over φ_2, $h_n(\varphi_1)$ can be obtained by integrating the r.h.s. of (14.27) over φ_2; e.g.

$$h_0(\varphi_1) = \int_0^{2\pi} \frac{d\varphi_2}{2\pi} f(\varphi_1\varphi_2) \quad.$$

(14.30)

In analogy with (13.6), (14.20) implies that

$$\frac{d}{dt} \int dx\, h_n(x, t; \varphi_1) = 0 \quad \forall_{n\in\mathbb{N}} \quad;$$

(14.31)

thus

$$Q_{nk} = \int dx \int \frac{d\varphi_1}{2\pi} e^{-ik\varphi_1} h_n(x, t; \varphi_1) \tag{14.32}$$

are the desired conserved quantities. For the first ($n = 0$) level one may check directly the time–independence, i.e. that of $\int dx \int d\varphi_2 \, f$ by using the equation of motion, (14.22). $\int \int g'$ is trivially zero, while verifying

$$\int \frac{d\varphi_2}{2\pi} (f * g - g * f) = 0 \tag{14.33}$$

requires (14.25) i.e. (crucially),

$$a_m(\varphi_1) = a_{-m}(\varphi_1 + m\lambda) \quad . \tag{14.34}$$

The determination of $h_1(\varphi_1)$ requires the knowledge of $r_1(\varphi)$. This can be obtained by comparing (14.25) with (14.24), which yields

$$r_1(\varphi) = \sum_{m \neq 0} \left(\frac{f_m(\varphi_1)}{c(\varphi_1) - c(\varphi_1 + m\lambda)} \right) * e^{im\varphi_2} \quad . \tag{14.35}$$

So

$$\begin{aligned} Q_1(\varphi_1) &= \int dx \, h_1(\varphi_1) = \int \frac{d\varphi}{2\pi} r_1 * f \\ &= \sum_{m \neq 0} \frac{f_m(\varphi_1) f_{-m}(\varphi_1 + m\lambda)}{c(\varphi_1) - c(\varphi_1 + m\lambda)} \quad . \end{aligned} \tag{14.36}$$

In order to have $Q_1(\varphi_1)$, and higher charges, non–singular, one is therefore required to take special c, e.g. $c = e^{ik\varphi_1}$.

Notes and References

(14.2) appears (implicitly) already in [1]. The three concrete (classes of) examples have divers sources; trigonometrical algebras were introduced in [2], their interpretation (14.6) and representation (14.9) in [3]. (14.10) goes back to [4], the non–isomorphy of different A_λ's is due to [5], the non–isomorphy of the corresponding Lie–Algebras was proven in [6]; concerning dynamical systems, one could note that the enveloping algebra of $sl(2)$ has played a role in the theory of relativistic membranes [7]. W_∞ was proposed in [8]; the uniqueness (and form) of the central extension of the Lie algebra L_0 of differential operators on S^1 is mentioned in [9] (a proof is given in [10]). The discussion of zero curvature type conditions on trigonometrical algebras is based on joint work with M. Olshanetsky and S. Theisen [11] — as well as on [12], which mainly concerns dynamical systems on A_λ and W_∞.

[1]   V.G. Drinfeld, V.V. Sokolov; Journal of Sov. Math. 30(1985)1975.

[2]   D.B. Fairlie, P. Fletcher, C. Zachos; Phys. Lett. B218 (89) 203.

[3]   J. Hoppe; Phys. Lett. B215 (88) 706, Int. J. Mod. Phys. A4 (89) 5235.

[4]   V. Bargmann; Annals of Math. Vol. 48#3 (1947) 568.

[5]   J. Dixmier; Journal of Algebra, 24 (1973) 551.

[6]   M. Bordemann, J. Hoppe, P. Schaller; Phys. Lett. B232 (89) 199.

[7]   J. Goldstone; unpublished.
      J. Hoppe; MIT Ph.D. Thesis, 1982.

[8]   C.N. Pope, L.J. Romans, X. Shen; Nucl. Phys. B339 (1990) 191.
      Phys. Lett. B242 (1990) 401.

[9]   A.O. Radul; JETP Lett. 50 (1989) 341.

[10] W.L. Li; Journal of Algebra 122 (1989) 64.

[11] J. Hoppe, S. Theisen, M. Olschanetsky; *Dynamical Systems on Quantum Tori Algebras*, KA–THEP–10/91.

[12] J. Hoppe; *Dynamical Systems on W_∞ and Universal Enveloping Algebras*, KA–THEP–11/91.

15 Spectral Transform and Solitons

The spectral transform technique as a method of solving certain classical non linear differential equations was introduced in [1], and has played an important role since then. Curiously, it uses the formulation (and abstract solution) of a problem known from quantum mechanics, namely that of solving

$$-\Psi''(x) + u(x)\Psi(x) = E\Psi(x) \quad (-\infty < x < +\infty) \tag{15.1}$$

for given $u(x)$, which is supposed to be a smooth real function and to go to zero (sufficiently fast) as $x \to \pm\infty$.

For $E = k^2 > 0$ there exist solutions characterized by

$$\Psi_k(x) \to \begin{cases} T(k)e^{-ikx} \,, & x \to -\infty \\ e^{-ikx} + R(k)e^{+ikx} \,, & x \to +\infty \end{cases} \tag{15.2}$$
$$- \infty < k < +\infty \;\; ;$$

$R(k)$ and $T(k)$ are called reflection–, and transmission coefficient respectively.

For particular (let us assume, finitely many) values of $E < 0$,

$$E_n = -\kappa_n^2 \quad \begin{matrix} \kappa_n > 0 \\ n = 1, 2, \ldots, N \geq 0 \end{matrix} \tag{15.3}$$

Eq. (15.1) has square–integrable real solutions Ψ_n ($n = 1, 2, \ldots N$). For $E = 0$ there will be one solution bounded at $\pm\infty$, but none square–integrable in the example(s) considered below. By the spectral transform $S[u(x)]$ one means the collection of

$$S[u] = \{R(k), -\infty < k < +\infty; \kappa_n, \rho_n | n = 1, 2, \ldots, N\} \,, \tag{15.4}$$

where the ρ_n are certain (positive) numbers characterizing the bound–state wave-functions, defined as

$$\rho_n = \left(\int_{-\infty}^{+\infty} \Psi_n^2(x)dx \right)^{-1} \,, \tag{15.5}$$

and the normalisation of the eigenfunctions Ψ_n is chosen such that

$$\lim_{x \to +\infty} \left(e^{\kappa_n x}\Psi_n(x) \right) = 1 \;\; . \tag{15.6}$$

The motivation for the above definition of $S[u]$ is that (for an appropriate function–class for u and certain conditions on $R(k)$) it provides a one–to–one correspondence: given the r.h.s. of (15.4), $u(x)$ is uniquely determined as the derivative of the solution to some inhomogenous Fredholm (specifically, Gel'fand–Levitan–Marchenko) integral equation (see [2] for a derivation and references to the original literature):

$$M(x) := \int_{-\infty}^{+\infty} \frac{dk}{2\pi} e^{ikx} R(k) + \sum_{1}^{N} \rho_n e^{-\kappa_n x} \tag{15.7}$$

$$K(x,y) + \int_{x}^{\infty} dz\, K(x,z) M(z+y) = -M(x+y) \quad y \geq x \tag{15.8}$$

$$u(x) := -2 \frac{\partial}{\partial x} K(x,x) \quad . \tag{15.9}$$

Let us look at the simplest non–trivial example, $R = 0$ and $N = 1$: In this case

$$M(x) = \rho e^{-\kappa x} \tag{15.10}$$

and (15.8), i.e.

$$K(x,y) = -\rho e^{-\kappa y} \left(e^{-\kappa x} + \int_{x}^{\infty} K(x,z) e^{-\kappa z} dz \right) \tag{15.11}$$

becomes seperable:

$$K(x,y) = -\rho e^{-\kappa y} f(x) \tag{15.12}$$

yields

$$f(x) = \frac{e^{-\kappa x}}{1 + \frac{\rho}{2\kappa} e^{-2\kappa x}} \tag{15.13}$$

so that, according to (15.9),

$$u(x) = \frac{-4\kappa\rho}{\left(e^{\kappa x} + \frac{\rho}{2\kappa} e^{-\kappa x} \right)^2} = \frac{-2\kappa^2}{\cosh^2 \kappa(x - x_0)} \quad , \quad x_0 = \frac{1}{2\kappa} \ln \frac{\rho}{2\kappa} \quad . \tag{15.14}$$

This result can be checked by explicitly solving the Schrödinger equation (15.1) with the potential u given by (15.14) (see the addendum), finding indeed $R(k) = 0$, and one bound state at energy $E = -\kappa^2$.

Requiring $R(k) = 0$, and $N > 1$ bound states at energies $-\kappa_n^2$ (and normalisation constants ρ_n), $n = 1, \ldots, N$, (15.8) — being 'seperable of rank N' — is almost as easy to solve: the Ansatz

$$K(x,y) = -\sum_{1}^{N} \sqrt{\rho_n} e^{-\kappa_n y} f_n(x) \tag{15.15}$$

leads to

$$(1 + A(x)) \mathbf{f}(x) = \mathbf{a}(x)$$

$$A_{mn}(x) = \frac{\sqrt{\rho_m \rho_n}}{\kappa_m + \kappa_n} e^{-(\kappa_m + \kappa_n)x} \quad , \tag{15.16}$$

$$a_n(x) = \sqrt{\rho_n} e^{-\kappa_n x}$$

i.e. *algebraic* equations for the $f_n(x)$, $n = 1, \ldots, N$. Hence

$$u(x) = 2\partial_x \left(\mathbf{a}^{\mathrm{tr}}(x) \left(\mathbb{1} + A(x) \right)^{-1} \mathbf{a}(x) \right) \quad , \tag{15.17}$$

which can be rewritten as

$$\begin{aligned}
u(x) &= 2\partial_x \operatorname{Tr} \left(\mathbf{a}^{\mathrm{tr}}(x) \left(\mathbb{1} + A(x) \right)^{-1} \mathbf{a}(x) \right) \\
&= 2\partial_x \operatorname{Tr} \left(\mathbf{a} \cdot \mathbf{a}^{\mathrm{tr}} \cdot \left(\mathbb{1} + A \right)^{-1} \right) \\
&= -2\partial_x^2 \operatorname{Tr} \ln \left(\mathbb{1} + A(x) \right) \\
&= -2\partial_x^2 \ln(\tau(x)) \\
\tau(x) &= \det \left(\mathbb{1} + A(x) \right) \quad .
\end{aligned} \tag{15.18}$$

Analogously, one can construct $u(x) = V_N(x)$ *recursively*: let

$$H_{N-1} = -\partial_x^2 + V_{N-1}(x) = A_{N-1}(x) = A_{N-1} A_{N-1}^+ - \kappa_{N-1}^2 \quad , \tag{15.19}$$

$$A_{N-1} = \partial_x + g_{N-1}(x) \quad , \tag{15.20}$$

be a one-dimensional Hamiltonian with 'reflectionless' potential $V_{N-1}(x)$ ($\to 0$, for $|x| \to \infty$) possessing $N - 1$ bound–states at $-\kappa_{N-1}^2 < \dots < -\kappa_1^2$. In order to add one discrete eigenvalue $E_N = -\kappa_N^2$ ($< -\kappa_{N-1}^2$) to the spectrum of H_{N-1} (while keeping $R(k) = 0$ for all k) write

$$H_{N-1} = A_N^+ A_N - \kappa_N^2 \tag{15.21}$$

$$A_N = \partial_x + g_N(x) \quad . \tag{15.22}$$

$$H_N = A_N A_N^+ - \kappa_N^2 = -\partial_x^2 + V_N(x) \tag{15.23}$$

will then have the same spectrum as H_{N-1}, apart from the additional eigenvalue $-\kappa_N^2$ with corresponding eigenfunction

$$\Psi_N(x) = e^{\int_{x_0}^{x} g_N(y)dy} \tag{15.24}$$

(see the addendum, where the case $N = 1$, with $\kappa_1 = 1$, $\rho_1 = 2$ and the notation $H_1 = H$, $H_0 = \tilde{H}$ is treated in detail).

According to (15.19/20/22),

$$g_N^2 - g_N' - \kappa_N^2 = V_{N-1}(x) = g_{N-1}^2 + g_{N-1}' - \kappa_{N-1}^2 \quad . \tag{15.25}$$

Note that this implies

$$\begin{aligned}
V_N(x) &= g_N^2 + g_N' - \kappa_N^2 = \left(g_{N-1}^2 - g_{N-1}' - \kappa_N^2 \right) + 2g_N' \\
&= V_{N-1}(x) + 2g_N' \\
&= \dots = 2 \sum_{n=1}^{N} g_n'(x) = 2\partial_x^2 \ln \left(\prod_{n=1}^{N} \Psi_n \right) \quad .
\end{aligned} \tag{15.26}$$

In order to determine g_N (respectively Ψ_N) explicitly, substitute

$$g_N(x) = -\frac{\Phi'_N(x)}{\Phi_N(x)} \tag{15.27}$$

in (15.25), yielding

$$-\Phi''_N + V_{N-1}(x)\Phi_N(x) = -\kappa_N^2 \Phi_N \quad . \tag{15.28}$$

This is the Schrödinger equation for H_{N-1}, but for $E = -\kappa_N^2$ which does not correspond to any eigenvalue of H_{N-1} (nor lies in the continuous part of the spectrum of H_{N-1}). Nevertheless, due to the relation of H_{N-1} to the free Hamiltonian $H_0 = -\partial_x^2$, we know that

$$\Phi_N^\pm := A_{N-1}A_{N-2}\ldots A_1 e^{\pm \kappa_N x} = \left(\partial_x - \frac{\Phi'_{N-1}}{\Phi_{N-1}}\right) \cdots \left(\partial_x - \frac{\Phi'_1}{\Phi_1}\right) e^{\pm \kappa_N x} \tag{15.29}$$

are two independent solutions of (15.28). Both of them are different from zero (for any finite x), as if $\Phi_N(x_0) = 0$,

$$V_{N-1}^\pm(x) := \begin{cases} V_{N-1}(x) \text{ if } x \mp x_0 < 0 \\ \infty \text{ if } x \mp x_0 > 0 \end{cases} . \tag{15.30}$$

would be strictly bigger than $V_{N-1}(x)$, but have $\Phi_N^\pm(x)$ — when set identically zero in the region where $V_{N-1}^\pm$ is infinite — as a square–integrable eigenfunction of $H_{N-1}^\pm = -\partial_x^2 + V_{N-1}^\pm$ with energy $-\kappa_N^2$ (smaller than the lowest eigenvalue of H_{N-1}), which can not be true. Furthermore,

$$\lim_{x \to +\infty} \left(e^{-\kappa_n x}\Phi_N^+(x)\right) = \prod_{n=1}^N (\kappa_N - \kappa_n) > 0$$

$$\lim_{x \to -\infty} \left(e^{+\kappa_N x}\Phi_N^-(x)\right) = (-1)^{N-1} \prod_{n=1}^N (\kappa_N - \kappa_n) \quad , \tag{15.31}$$

so that

$$\Phi_N(x) = A_{N-1}\cdots A_1 \left(\overbrace{\sqrt{\alpha_N}e^{\kappa_n x} + \frac{(-1)^{N-1}}{\sqrt{\alpha_N}}e^{-\kappa_N x}}^{=:u_N(x)}\right) \tag{15.32}$$

will be strictly positive for all $\alpha_N > 0$. The reason for caring so much about the positivity of Φ_N is that only then (15.27) and

$$V_N(x) = -2\partial_x^2 \ln \left(\prod_1^N \Phi_n(x)\right) \tag{15.33}$$

are well defined; also, $\Psi_N \sim \frac{1}{\Phi_N}$ should be square integrable, which is why the cases where Φ_N is proportional to a pure exponential, $\alpha_N \hat{=} 0$ and $\alpha_N \hat{=} \infty$, have to be excluded as well (which has already been assumed in (15.31)).

Now that we have (recursively) found $V_N(x) = V_N(x, \alpha_1, \ldots, \alpha_N)$, with exactly N bound states, prescribed at energies $-\kappa_1^2, \ldots, -\kappa_N^2$, it is interesting to see whether one can convert this recursive procedure into a direct formula for V_N (of course, we already know an answer, as we have solved the GLM integral equations, cp. equation (15.18), but it is worthwhile to obtain an answer by *direct* methods). To get an idea, let us evaluate Φ_N and

$$\chi_N = \prod_1^N \Phi_n = \Phi_N \chi_{N-1} \tag{15.34}$$

for $N = 1$, and $N = 2$:

$$\Phi_1 = \cosh \kappa_1 (x - \tilde{x}_1) = \chi_1$$
$$g_1 = -\kappa_1 \tanh \kappa_1 (x - \tilde{x}_1)$$
$$\Phi_2 = (\partial_x - \kappa_1 \tanh(1)) \sinh \kappa_2 (x - \tilde{x}_2)$$
$$\quad = \kappa_2 \cosh(2) - \kappa_1 \tanh(1) \sinh(2)$$
$$\chi_2 = \kappa_2 \cosh(2) \cosh(1) - \kappa_1 \sinh(1) \sinh(2) \tag{15.35}$$
$$g_2 = -\frac{\Phi_2'}{\Phi_2} = -\frac{\Phi_1 \Phi_2'}{\chi_2}$$
$$\quad = -\frac{1}{\chi_2} \left(\kappa_2^2 \sinh(2) \cosh(2) - \kappa_1 \kappa_2 \cosh(2) \sinh(1) - \kappa_1^2 \frac{\sinh(2)}{\cosh(1)} \right) \quad .$$

While Φ_2 and g_2 already contain hyperbolic functions in the denominator, χ_1 and χ_2 are polynominals in $\cosh(1$ or $2)$ and $\sinh(1$ or $2)$, i.e. $u_1(x)$, $u_2(x)$ and their derivatives. Note that we have put

$$\alpha_n = e^{2\kappa_n \tilde{x}_n} \tag{15.36}$$

so that

$$u_n(x) = \begin{cases} \cosh \kappa_n(x - \tilde{x}_n) \text{ for } n \text{ odd} \\ \sinh \kappa_n(x - \tilde{x}_n) \text{ for } n \text{ even} \end{cases} \quad . \tag{15.37}$$

Remarkably,

$$\chi_n(x) = \det \begin{pmatrix} u_1 & u_2 & \cdots & u_n \\ u_1' & u_2' & & u_n' \\ \vdots & \vdots & \ddots & \vdots \\ u_1^{(n-1)} & u_2^{(n-1)} & \cdots & u_n^{(n-1)} \end{pmatrix} =: W_n(u_1, \ldots, u_n) \tag{15.38}$$

for *all* $n \in \mathbb{N}$. To prove this we need a classical theorem of ordinary differential equations ('Jacobi's theorem'), stating that if $W_n \neq 0$ (and $W_0 \equiv 1$),

$$W_n(u_1, \ldots, u_n) W_n'(u_1, \ldots, u_{n-1}, u_{n+1})$$
$$- W_n(u_1, \ldots, u_{n-1}, u_{n+1}) W_n'(u_1, \ldots, u_n) \tag{15.39}$$
$$= W_{n-1}(u_1, \ldots, u_{n-1}) W_{n+1}(u_1, \ldots, u_n, u_{n+1}) \quad .$$

Using (15.39), the proof of (15.38) is easy: Suppose, (15.38) holds for all $m \leq n$, in particular

$$\Phi_m(x) = \frac{W_m(u_1,\ldots,u_m)}{W_{m-1}(u_1,\ldots,u_{m-1})} = \Phi_m[u_1,\ldots,u_m] \quad . \tag{15.40}$$

This implies that, according to (15.32),

$$
\begin{aligned}
\Phi_{n+1}(x) &= \left(\partial_x - \frac{\Phi_n'[u_1,\ldots,u_n]}{\Phi_n[u_1,\ldots,u_n]}\right)\Phi_n[u_1\ldots u_{n-1}u_{n+1}] \\
&= \left(\partial_x - \frac{W_n'(u_1,\ldots,u_n)}{W_n} + \frac{W_{n-1}'}{W_{n-1}}\right)\left(\frac{W_n(u_1,\ldots,u_{n-1},u_{n+1})}{W_{n-1}(u_1,\ldots,u_{n-1})}\right) \\
&= \frac{1}{W_{n-1}}\left(\partial_x \tilde{W}_n - \frac{W_n'}{W_n}\tilde{W}_n\right) \qquad \tilde{W}_n = W_n(u_1,\ldots,u_{n-1},u_{n+1}) \\
&= \frac{1}{W_{n-1}}\left(W_{n+1}\frac{W_{n-1}}{W_n}\right) = \frac{W_{n+1}(u_1,\ldots,u_{n+1})}{W_n(u_1,\ldots,u_n)} \quad .
\end{aligned}
$$

$$\tag{15.41}$$

Now we have 'two' formulae for $u(x) = V_N(x) = -2\partial_x^2 \ln f(x)$ ((15.18) and (15.33) with $f(x) = \prod_1^N \Phi_n = \chi_N$ given by (15.38)). This does not necessarily imply the coincidence of $\tau(x) := \det(\mathbb{1} + A(x))$ and $W(u_1,\ldots,u_N)$. According to $(\ln \tau)'' = (\ln W)''$ one may have $W(x) = ce^{bx}\tau(x)$, resulting in identical potential–functions $u(x)$. c and b may be deduced from the asymptotic behaviour of W and τ; comparing at $x \to +\infty$ ($\tau(x) \to 1$), one finds

$$W_N(x) = \frac{D(\kappa)}{2^N} e^{-\sum_1^N \kappa_n \tilde{x}_n} e^{\sum_1^N \kappa_n x} \cdot \tau(x)$$

$$D(\kappa) = \begin{vmatrix} 1 & 1 & \cdots & 1 \\ \kappa_1 & \kappa_2 & \cdots & \kappa_N \\ \vdots & \vdots & \ddots & \vdots \\ \kappa_1^{N-1} & \kappa_2^{N-1} & \cdots & \kappa_N^{N-1} \end{vmatrix} = \prod_{1 \le i < j \le N}(\kappa_j - \kappa_i) \quad , \tag{15.42}$$

where the constants $\tilde{x}_n$ and ρ_m (cp. (15.16)) are non-trivially related to each other. Finally, let us state yet another formula for $\tau(x)$ (see e.g. [3]):

$$\tau(x) = \sum_{\mu_n=0,1}\left\{e^{-2\sum_1^N \mu_n \kappa_n(x-x_n)}\prod_{1 \le i < j \le N}\left(\frac{\kappa_i - \kappa_j}{\kappa_i + \kappa_j}\right)^{2\mu_i\mu_j}\right\} \quad . \tag{15.43}$$

$$x_n = \frac{1}{2\kappa_n}\ln\frac{\rho_n}{2\kappa_n}$$

Let us now go back to our original goal, the solution of classical non–linear (partial) differential equations via the spectral transform.

Suppose, $u(x)$ depends on some external parameter t (which in the 'quantum–mechanical' consideration has certainly nothing to do with 'time'). Then $S[u(x,t)]$, the spectral transform, will of course also depend on t, i.e. $S = S(t)$. The most important fact is that for a large class of *non* linear evolution equations for $u = u(x,t)$, where t is 'time', $S(t)$ satisfies *linear* differential equations (which means that $u(x,t)$ may be found by transforming $u(x,0)$ into $S(0)$, solving for $S(t)$, and then transforming back to $u(x,t)$). Specifically, for any polynominal P, $S(t)$ develops according to

$$\dot{R}(k,t) = 2ikP(-4k^2)R(k,t)$$
$$\dot{\kappa}_n(t) = 0 \qquad\qquad\qquad (15.44)$$
$$\dot{\rho}_n(t) = -2\kappa_n P(4\kappa_n^2)\rho_n(t)$$

if $u(x,t)$ develops according to

$$\dot{u} = P(T)u' \quad , \qquad\qquad\qquad (15.45)$$

with $T = T[u]$ being an integro–differential operator defined as

$$(Tf)(x) := f''(x) - 4uf + 2u' \int_x^\infty dy\ f(y) \quad . \qquad\qquad (15.46)$$

A crucial feature of (15.44) is the time–independence of all κ_n.

As the simplest non–trivial example, one should note the case $P(T) = -T$, yielding the KdV equation for u,

$$\dot{u} + u''' - 6uu' = 0 \qquad\qquad\qquad (15.46)$$

and

$$R(k,t) = R(k,0)e^{8ik^3 t}$$
$$\rho_n(t) = \rho_n(0)e^{8\kappa_n^3 t} \qquad n = 1,\dots,N \qquad\qquad (15.48)$$
$$\kappa_n(t) = \kappa_n(0) = \kappa_n$$

as the solution of (15.44). This means that if u(x,0) is given by (15.18) (or (15.38) or (15.43)), the corresponding solutions $u(x,t)$ of (15.47) can be obtained by simply replacing ρ_i by $\rho_i(t)$ in the static formula (15.18), or substituting

$$x_i(t) = x_i + \kappa_i^2 t \qquad\qquad\qquad (15.49)$$

for x_i, in formula (15.43). In particular, (15.12) becomes the famous '1–soliton'–solution of (15.46)

$$u(x,t) = \frac{-2\kappa^2}{\cosh^2 \kappa(x - x_1 - 4\kappa^2 t)} \quad . \qquad\qquad (15.50)$$

For $N > 1$, only the asymptotic behaviour $(t \to \pm\infty)$ becomes simple,

$$u_N(x, t \to \pm\infty) \approx -2 \sum_{n=1}^N \frac{\kappa_n^2}{\cosh^2\left(\kappa_n\left(x - z_n^\pm - v_n t\right)\right)} \quad , v_n = 4\kappa_n^2 \quad , \qquad (15.51)$$

which is a sum of N solitons of the form (15.50), each moving to the right with fixed velocity, and shape (fast, high and thin if κ is large, while slow, low and broad if κ is small).

The shift (relative to what the corresponding constant x_n would have been without interaction)

$$\Delta_n = z_n^+ - z_n^- \qquad\qquad\qquad (15.52)$$

is given by

$$\kappa_n \Delta_n = \sum_{i=1}^{n-1} \Delta(\kappa_n, \kappa_i) - \sum_{j=n+1}^{N} \Delta(\kappa_n, \kappa_j)$$

$$\Delta(\kappa_i, \kappa_j) = \ln \left| \frac{\kappa_i + \kappa_j}{\kappa_i - \kappa_j} \right|$$

(15.53)

and can be interpreted as the gain (or loss) due to overtaking and being over-taken; in each individual 'scattering process' the faster soliton is pushed forward additionally, while the slower one is delayed. The remaining N constants (those *not* determined by (15.52/53)) correspond to the N ρ's (respectively x's,α's or $\tilde{x}$'s). Finally one should at least mention the ubiquitous importance of the τ function, having appeared here as

$$u(x, t) = -2 \partial_x^2 \ln \tau(x, t) \quad .$$

(15.54)

Let us e.g. write (15.47) in terms of τ, by letting $u = -w'$, integrating once (with a particular choice of integration constant),

$$\frac{1}{2} \dot{w} + \frac{3}{2} w'^2 + \frac{1}{2} w''' = 0 \quad ,$$

(15.55)

and then calculating derivatives of $\frac{1}{2} w = \frac{\tau'}{\tau}$:

$$\frac{1}{2} \dot{w} = \frac{\tau \dot{\tau}_x - \dot{\tau} \tau_x}{\tau^2} \quad ,$$

$$\frac{1}{2} w' = \frac{\tau \tau_{xx} - \tau_x^2}{\tau^2} = \frac{\tau_{xx}}{\tau} - \frac{\tau_x^2}{\tau^2} \quad ,$$

$$\frac{1}{2} w'' = \frac{\tau_{xxx}}{\tau} - \frac{3 \tau_x \tau_{xx}}{\tau^2} + 2 \frac{\tau_x^3}{\tau^3} \quad ,$$

$$\frac{1}{2} w''' = \frac{\tau_{xxxx}}{\tau} - \frac{4 \tau_x \tau_{xxx}}{\tau^2} - \frac{3 \tau_{xx}^2}{\tau^2} + \frac{12 \tau_x^2 \tau_{xx}}{\tau^3} - \frac{6 \tau_x^4}{\tau^4} \quad ,$$

$$\frac{3}{2} w'^2 = 6 \frac{\tau_{xx}^2}{\tau^2} - 12 \frac{\tau_x^2 \tau_{xx}}{\tau^3} + 6 \frac{\tau_x^4}{\tau^4} \quad .$$

(15.56)

Adding the last two equalities, all terms cubic (and higher) in τ and its derivatives cancel, and one obtains as (15.55) a purly quadratic (differential) equation for τ,

$$\dot{\tau}' \tau - \dot{\tau} \tau' + 3 \tau''^2 - 4 \tau' \tau''' + \tau'''' \tau = 0 \quad ,$$

(15.57)

which will (more or less) be the starting point of the next lecture.

Addendum: A Simple Quantum Mechanical Problem

Let us consider the time–independent Schrödinger equation

$$\left(-\partial_x^2 + V(x)\right)\Psi(x) = E\Psi(x) \tag{A.1}$$

for a parcticle moving in the one dimensional potential

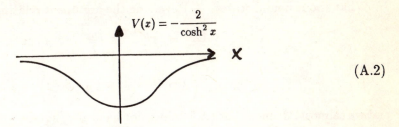

$$V(x) = -\frac{2}{\cosh^2 x} \tag{A.2}$$

This potential being one of the few known cases for which (A.1) can be solved explicitely (traditionally, this is done by expressing Ψ in terms of hypergeometric functions), the solutions may also be obtained by noting that

$$H = -\partial_x^2 + V(x) = AA^+ - 1 \quad , \tag{A.3}$$

$$A = \partial_x - \tanh x \quad , \tag{A.4}$$

is related to the Hamiltonian of a free particle,

$$\tilde{H} = -\partial_x^2 = A^+ A - 1 \quad . \tag{A.5}$$

As

$$(A^+ A - 1)\tilde{\Psi}^{(E)} = E\tilde{\Psi}^{(E)} \tag{A.6}$$

implies

$$(AA^+ - 1)(A\tilde{\Psi}^{(E)}) = E(A\tilde{\Psi}^{(E)}) \tag{A.7}$$

and

$$\tilde{\Psi}_k = e^{-ikx} \quad , \quad -\infty < k < \infty \tag{A.8}$$

solves (A.6) with

$$E = E(k) = k^2 \geq 0 \quad , \tag{A.9}$$

the wavefunction

$$\Psi_k(x) := A\tilde{\Psi}_k(x) = -(ik + \tanh x)e^{-ikx} \tag{A.10}$$

solves (A.1), with $E = E(k)$ given by (A.9). Defining reflection– and transmission coefficients according to

$$\Psi_k(x) \rightarrow \begin{cases} e^{-ikx} + R(k)e^{+ikx} & , \quad x \rightarrow +\infty \\ T(k)e^{-ikx} & , \quad x \rightarrow -\infty \end{cases} \tag{A.11}$$

we find, with $\bar{\Psi} = \frac{-\Psi_k}{1+ik}$,

$$R(k) = 0 \quad , \quad T(k) = \frac{k+i}{k-i} \quad ; \tag{A.12}$$

they satisfy

$$|R(k)|^2 + |T(k)|^2 = 1$$
$$T^*(k) = T(-k) \quad , \quad R^*(k) = R(-k) \tag{A.13}$$

— as they should.

What about bound states? — Reversing the argument relating (A.6) and (A.7),

$$(AA^+ - 1)\Psi^{(E)} = E\Psi^{(E)} \tag{A.7'}$$

implies

$$(A^+ A - 1)\left(A^+ \Psi^{(E)}\right) = E\left(A^+ \Psi^{(E)}\right) \quad , \tag{A.6'}$$

and we (almost) deduce that (A.7') does not have any (square–integrable) solution for $E \leq 0$, as the free Schrödinger equation doesn't. There is one case for which the above argument does not hold, namely

$$E = -1 \quad , \quad A^+ \Psi^{(0)} = 0 \quad , \tag{A.14}$$

and $\Psi^{(0)}$ square–integrable. Indeed, (A.14) does have a normalizable solution

$$\Psi^{(0)}(x) = \frac{c}{\cosh x} \quad , \tag{A.15}$$

which is easily found by integrating

$$(\ln \Psi(x))' = -\tanh x \quad ; \tag{A.16}$$

as

$$\int_0^\infty \frac{1}{\cosh^2 x} = \tanh x\big|_0^\infty = 1 \quad , \tag{A.17}$$

$c = 1/\sqrt{2}$ will make $\int |\Psi|^2 = 1$.

As the case $E = 0$, the lower boundary of the continuous spectrum, is rather special (in [7] it is — falsely — listed as a bound state energy and claimed to be an eigenvalue of $H = -\partial_x^2 - \frac{2}{\cosh^2 x}$) it may be interesting to give three additional arguments that there can't exist a square–integrable solution of (A.1) when $E = 0$.
A) Consider

$$\chi(x) = \int_{-\infty}^{+\infty} D(x, x')\Psi(x')dx' \quad , \tag{A.18}$$

$$D(x, x') = \delta(x - x') - \int_{-\infty}^{+\infty} \frac{dp}{p^2 + 1}\Psi_p(x)\Psi_p^*(x') \quad . \tag{A.19}$$

Using (A.10), it is not difficult to check that, independent of Ψ,

$$A^+ \chi(x) = 0 \quad . \tag{A.20}$$

This means that $\chi(x)$ must be proportional to $\Psi^{(0)}$, and $D(x, x')$ — being the missing part in the completeness relation,

$$\delta(x - x') = \int dp\mu(p)\Psi_p(x)\Psi_p^*(x') + \sum_{n=0}^{N-1} \Psi^{(E_n)}(x)\Psi^{(E_n)^*}(x') \quad , \qquad (A.21)$$

is simply equal to $\Psi^{(0)}(x)\Psi^{(0)^*}(x')$, and $N = 1$.

B) $E = 0$, with V as in (A.2), is a special case of a 'coupling–constant threshold'; a general theorem [8] states that for one dimensional 'short–range' potentials there can *not* exist a normalizable state at the border of the continuum (strictly speaking, 'short–range' is defined in [8] as of compact support, but the same theorem should hold for exponentially decaying potentials).

C) The most direct proof that there can't be a bound state with energy 0 is, of course, to simply solve (A.1). As we already know one solution ($E = 0$),

$$\Psi_{k=0}(x) = -\tanh x = \Psi_1(x) \quad , \qquad (A.22)$$

one may try directly the Ansatz $\Psi(x) = f(x)\tanh x;$ — or first transform (A.1) with $E = 0$ by putting

$$\Psi(x) = \frac{w(\sinh^2 x)}{\cosh x} \quad , \qquad (A.23)$$

and then solve the corresponding (hypergeometric) equation,

$$z(1 + z)w''(z) + \frac{1}{2}w'(z) + \frac{1}{4}w(z) = 0 \qquad z \geq 0 \quad . \qquad (A.24)$$

The Ansatz z^α gives a characteristic equation for α both roots of which are $= \frac{1}{2}$. So

$$w_1(z) = \sqrt{z} \qquad (A.25)$$

(leading to (A.22)); the 'variation of constant' Ansatz $w_2(z) = \sqrt{z}C(z)$ leads to

$$C'(z) = \frac{\sqrt{1 + z}}{z^{\frac{3}{2}}} \qquad (A.26)$$

which can be integrated,

$$C(z) = C - 2\sqrt{\frac{1 + z}{z}} + \ln\left(1 + 2z + 2\sqrt{z(1 + z)}\right) \quad . \qquad (A.27)$$

Using (A.23), one finds

$$\Psi(x) = x\tanh x - 1 + C \cdot \tanh x \quad . \qquad (A.28)$$

Neither (A.22) nor (A.28) is normalizable.

Notes and References

The GLM integral equation, as well as (15.44/45) and (15.49/51) are taken from [2], (15.43) from [3]. The addendum is mostly based on [4]. The explicit determination (using Wronskians) of potentials that give rise to n bound states was already discussed (for a finite interval, instead of the whole line) in [5] (see also [6] and references therein). Concerning Bäcklund transformations, Vertex operators (furnishing the transition from τ_N to τ_M), and much more beautiful structure related to the spectral transform and solitons (as well as for references on the original literature) see [2] and [3].

[1]   C.S. Gardener, J.M. Greene, M.D. Kruskal, R.M. Miura; Phys. Rev. Lett. 19 (1967) 1095

[2]   F. Calogero, A. Degasperis; *Solitons and the Spectral Transform I*, North Holland 1982.

[3]   A.C. Newell; *Solitons in Mathematics and Physics*, CBMS-NSF Ref. Conf. Ser. in Appl. Math. 48 SIAM 1985.

[4]   J. Goldstone; *Solution to Problem Set #1*, Quantum Mechanics Course, MIT 1977.

[5]   M.M. Crum; Quart. J. Math. 6 (1955) 121.

[6]   P. Deift, E. Trubowitz; Comm. Pure and Appl. Math. Vol. 32 (1979) 121.

[7]   S. Flügge; *Practical Quantum Mechanics I*, Problem 39 (p.100), Springer 1971.

[8]   M. Klaus, B. Simon; Annals of Physics 130 (1980) 251.

16 Higher Dimensional τ–Functions

The notion of τ–function, discovered by Hirota [1] about twenty years ago, is central to the theory of integrable systems. In particular, it allows explicit formulae for multi–soliton solutions of the 1+1 dimensional non–linear equations of the KdV (Korteweg/de Vries)–hierarchy and, in the generalized form of determinants of infinite–dimensional matrices [2] simultaneous solutions of the 2+1 dimensional KP (Kadomtsev/Petviashvili) systems. In both cases τ depends on infinitely many 'time' variables, while it is the second derivative (of $\ln(\tau)$) with respect to *one* variable, x, which enters the original non–linear equations in their conventional form.

In any case, Hirota's bilinear equations are formulated in terms of operators $D_x, D_t, \ldots$, each of which involves only *one* variable at a time, and it would be interesting to find intrinsically higher dimensional formulations of $2+1$, $3+1, \ldots$ dimensional systems possessing soliton solutions.

Hirota's 'antisymmetric' differentiation of products of functions,

$$D_x(f \cdot g) = \lim_{\epsilon \to 0} \frac{\partial}{\partial \epsilon} \left(f(x+\epsilon) g(x-\epsilon) \right) = f_x g - f g_x \tag{16.1}$$

takes into account the signs in

$$\tau \tau_{xt} - \tau_x \tau_t + \tau \tau_{xxxx} - 4\tau_x \tau_{xxx} + 3\tau_{xx} \tau_{xx} = 0 \quad , \tag{16.2}$$

which is the form the KdV equation

$$u_t - 6uu_x + u_{xxx} = 0 \tag{16.3}$$

takes upon writing

$$u(x,t) = -2\partial_x^2 \ln \tau(x,t) \quad . \tag{16.4}$$

Using (16.1), Eq. (16.2) can be written as

$$\left(D_x D_t + D_x^4 \right) \tau \cdot \tau = 0 \quad , \tag{16.5}$$

the quantity inside the bracket being the simplest, and most famous, 'Hirota-polynomial' admitting N soliton solutions of the form [3]

$$\tau_N = \sum_{\mu_j = 0,1} e^{\left(\sum_{i=1}^{N} \mu_i H_i + \sum_{1 \le i < j \le n} A_{ij} \mu_i \mu_j \right)}$$

$$0 < \kappa_1 < \ldots < \kappa_N \quad ; \quad e^{A_{ij}} = \left(\frac{\kappa_i - \kappa_j}{\kappa_i + \kappa_j} \right)^2 \tag{16.6}$$

— the H_j are linear functions of x and t (and all other 'times' entering the equations of the KdV hierarchy).

The three simplest non–trivial ($\tau \neq 1$) solutions of (16.5) are

$$\tau_1 = 1 + e^{H_1}$$

$$\tau_2 = 1 + e^{H_1} + e^{H_2} + \left(\frac{\kappa_1 - \kappa_2}{\kappa_1 + \kappa_2}\right)^2 e^{H_1 + H_2}$$

$$\tau_3 = 1 + \sum_{i=1}^{3} e^{H_i} + \sum_{1 \leq i < j \leq 3} e^{H_i + H_j + A_{ij}} + e^{H_1 + H_2 + H_3 + A_{12} + A_{23} + A_{31}}$$

$$H_i = \kappa_i \left(x - x_i - 4\kappa_i^2 t\right) \quad .$$

(16.7)

The corresponding solutions of (16.3) asymptotically describe N $(= 1, 2, 3 \ldots)$ seperated solitons,

$$u(x, t \to \pm\infty) \approx \sum_{i=1}^{N} \frac{-2\kappa_i^2}{\cosh^2 \kappa_i \left(x - x_i^{\pm} - 4\kappa_i^3 t\right)} \quad , \tag{16.8}$$

where the $x_i^{\pm}$ are real constants that depend on the x_i and κ_i. Finding (or verifying) these solutions is very easy when using the fact that products of exponentials are 'eigenfunctions' of the Hirota D–operators,

$$D_x \left(e^{mx} \cdot e^{nx}\right) = (m - n) e^{(m+n)x} \quad m, n \in \mathbb{C} \quad . \tag{16.9}$$

If one defines the commutator $[\ ,\ ]$ of $L_m = e^{mx}$ and $L_n = e^{nx}$ to be $D_x(L_m L_n)$ one trivially obtains

$$[L_m, L_n] = (m - n)L_{m+n} \quad . \tag{16.10}$$

Of course, there are much more profound manifestations of infinite dimensional algebras in the theory of τ–functions. Nevertheless, (16.9/10) naturally suggests what kind of 'non–commutative' operators (acting on products of functions) might play a role for higher dimensional equations. Replacing (16.10) by

$$[T_{\mathbf{m}}, T_{\mathbf{n}}] = (\mathbf{m} \times \mathbf{n})T_{\mathbf{m+n}}$$

$$\mathbf{m} \times \mathbf{n} = m_1 n_2 - m_2 n_1 \quad ; \quad \mathbf{m}, \mathbf{n} \in \mathbb{C}^2$$

(16.11)

one would look for an operator $D_0 = D_x$ satisfying

$$D_0 \left(e^{\mathbf{m}\mathbf{x}} \cdot e^{\mathbf{n}\mathbf{x}}\right) = (\mathbf{m} \times \mathbf{n})e^{(\mathbf{m+n})\mathbf{x}} \quad . \tag{16.12}$$

This will be the case if one defines

$$(D_0 (f \cdot g))(\mathbf{x}) = \epsilon_{rs} \partial_r f(\mathbf{x}) \partial_s f(\mathbf{x}) = \{f, g\}(\mathbf{x})$$

$$= \lim_{\mathbf{x}' \to \mathbf{x}} \left((\epsilon_{rs} \partial_r \partial_s') f(\mathbf{x})g(\mathbf{x}')\right) \quad . \tag{16.13}$$

Unlike (16.10), (16.11) admits a one–parameter Lie algebra deformation, the trigonometrical algebras [4],

$$[T_{\mathbf{m}}, T_{\mathbf{n}}] = \frac{1}{2\pi\Lambda}\sin\left(2\pi\Lambda(\mathbf{m}\times\mathbf{n})\right)T_{\mathbf{m}+\mathbf{n}} \quad \Lambda\in\mathbb{C} \quad . \tag{16.14}$$

Correspondingly, one could try to (define and) interpret

$$\begin{aligned}
(D_\lambda(f\cdot g))(\mathbf{x}) &= \frac{1}{\lambda}\left(f * g - g * f\right)(\mathbf{x}) \\
&= \frac{1}{\lambda}\lim_{\mathbf{x}'\to\mathbf{x}}\left(e^{\frac{\lambda}{2}\epsilon_{rs}\partial_r\partial_s'}\left(f(\mathbf{x})g(\mathbf{x}') - g(\mathbf{x})f(\mathbf{x}')\right)\right) \quad \lambda\in\mathbb{C}
\end{aligned} \tag{16.15}$$

as leading to 'quantized' versions of classical non–linear equations $(D_{\lambda\to0}(f\cdot g) = \lim_{\lambda\to0}\left(\frac{1}{\lambda}[f,g]^*\right) = \{f,g\} = D_0)$.

Also, one may easily generalize (16.13) to other 'topologies', as well as to yet higher dimensions:

$$(D_A(f\cdot g))(\mathbf{x}) = \sum_{r,s=1}^{d>2} A_{rs}\partial_r f(\mathbf{x})\partial_s g(\mathbf{x}) \tag{16.16}$$

$$A_{rs} = -A_{sr}\in\mathbb{C} \quad ,$$

corresponding to the algebras

$$[T_{\mathbf{m}}, T_{\mathbf{n}}] = \left(\mathbf{m}^{\mathrm{tr}}A\mathbf{n}\right)T_{\mathbf{m}+\mathbf{n}} \quad A^{\mathrm{tr}} = -A \, ; \, \mathbf{m}, \mathbf{n}\in\mathbb{C}^d \quad . \tag{16.17}$$

In any case, when trying to find out whether equations like $(D_0 D_t + D_0^{2k})J\cdot J = 0$ (I will write J for the higher dimensional τ–functions) can have solutions analogous to (16.6), i.e. (weighted) sums of exponentials $e^{\mathbf{mx}-a_m t}$, one finds that (for $N>2$) either all $\mathbf{m}$ have to be parallel to each other (trivially leading to solutions) or ('quasi one dimensional' case) have to 'end' on a straight line, $\mathbf{m}_i - \mathbf{m}_j = (m_i - m_j)\mathbf{a}$. Morover, if one wants $u(\mathbf{x},t) = -2\nabla^2\ln(J(\mathbf{x},t))$ to be the generalization of (16.4), the equations for J should be manifestly invariant under multiplication of J by $c\cdot e^{\mathbf{c}\cdot\mathbf{x}}$. This will necessarily be the case, if one defines

$$\begin{aligned}
\left(D^k(f\cdot g\cdot h)\right)(\mathbf{x}) &= \lim_{\mathbf{x}^{(\alpha)}\to\mathbf{x}}\left((D_{12} + D_{23} + D_{31})^k f(\mathbf{x}^{(1)})g(\mathbf{x}^{(2)})h(\mathbf{x}^{(3)})\right) \\
D_{\alpha\beta}\left(f(\mathbf{x}^{(1)})g(\mathbf{x}^{(2)})h(\mathbf{x}^{(3)})\right) &= \epsilon_{rs}\partial_r^{(\alpha)}\partial_s^{(\beta)}\left(f(\mathbf{x}^{(1)})g(\mathbf{x}^{(2)})h(\mathbf{x}^{(3)})\right) \quad ,
\end{aligned} \tag{16.18}$$

where ∂_r^α denotes the derivative with respect to $x_r^{(\alpha)}$. Explicitly,

$$\begin{aligned}
D(f\cdot g\cdot h) &= f\{g,h\} + g\{h,f\} + h\{f,g\} \\
D^2(f\cdot g\cdot h) &= (f_{11}g_{22} + f_{22}g_{11} - 2f_{12}g_{12})h + \text{cycl.} \\
&\quad + 2\cdot(f_1 h_{12}g_2 + f_2 h_{12}g_1 - f_1 h_{22}g_1 - f_2 h_{11}g_2) + \text{cycl.} \quad .
\end{aligned} \tag{16.19}$$

Analogous to (16.15), one may define the $*$ product quantization of (16.18). Note that $D(f\cdot g\cdot h)$ is totally antisymmetric in f, g and h and that the vanishing of any polynomial in D, acting on $J\cdot J\cdot J$, will be invariant under $J\to ce^{\mathbf{c}\cdot\mathbf{x}}J$ as

$$D(\tilde{f}_1\cdot\tilde{f}_2\cdot\tilde{f}_3) = c^3 e^{3\mathbf{c}\mathbf{x}}D(f_1\cdot f_2\cdot f_3) \quad \text{if } \tilde{f}_i = ce^{\mathbf{c}\mathbf{x}}f_i \quad . \tag{16.20}$$

Furthermore,

$$D^k(e^{mx} \cdot e^{nx} \cdot e^{kx}) = \left((m \times n) + (n \times k) + (k \times m)\right)^k e^{(m+n+k)x} \quad , \quad (16.21)$$

so that the powers of D, acting on $J \cdot J \cdot J$ will be easy to evaluate for J being a sum of exponentials. What shall we do with D_t? Obviously, one needs to generalize Hirota's bilinear D_t-operator to a trilinear one. Let us take

$$D_t(f \cdot g \cdot h) = \det \begin{pmatrix} f & g & h \\ \dot{f} & \dot{g} & \dot{h} \\ \ddot{f} & \ddot{g} & \ddot{h} \end{pmatrix} = lim_{t_\alpha \to t} \left((d_{12} + d_{23} + d_{31})\, f(t_1)g(t_2)f(t_3)\right)$$

$$d_{\alpha\beta} := \partial_{t_\alpha} \partial_{t_\beta}^2 - \partial_{t_\beta} \partial_{t_\alpha}^2 \quad ,$$

$$(16.22)$$

which is linear in the three arguments, and totally antisymmetric.

Consider then an (admittedly oversimplified) example,

$$(D_t D + D^2)\, J \cdot J \cdot J = 0 \quad . \tag{16.23}$$

As

$$(D_t D + D^2) \left(e^{m_1 x - a_1 t} \cdot e^{m_2 x - a_2 t} \cdot e^{m_3 x - a_3 t}\right)$$

$$= ((m_1 \times m_2) + (m_2 \times m_3) + (m_3 \times m_1))\, e^{\sum_{i=1}^{3}(m_i x - a_i t)}.$$

$$\cdot \Big\{ a_1 a_2 (a_1 - a_2) + a_2 a_2 (a_2 - a_3) + a_3 a_1 (a_3 - a_1) + \tag{16.24}$$

$$+ (m_1 \times m_2) + (m_2 \times m_3) + (m_3 \times m_1) \Big\} \quad ,$$

Eq. (16.23) will 'trivially' (i.e. with no amplitudes to be arranged, and no shifts in the transition from $-\infty$ to ∞) hold for

$$J = \sum_{i=1}^{N} e^{(m_i x - a_i t)} \tag{16.25}$$

(one may also consider infinite sums, or integrals) *provided*

$$a_i a_j (a_j - a_i) = m_i \times m_j \quad \forall_{ij} \quad . \tag{16.26}$$

Dividing by $a_i a_j$ one sees that the areas $F_{ij} := (m_i/a_i) \times (m_j/a_j)$ have to satify $F_{ij} + F_{jk} = F_{ik}$, which means that the m_i/a_i have to 'end' all on one line. So

$$m_i = \left(\kappa_i, \kappa_i^2\right) \quad , \quad a_i = \kappa_i \quad . \tag{16.27}$$

One may contrast this with the 'multi-soliton' solutions [6],

$$v(x, t) = \kappa^2 - 2\nabla^2 \ln \int_0^{2\pi} f(\theta) e^{\hat{\kappa} \cdot x + 2\kappa^3 t \cos 3\theta} d\theta$$

$$\left(\hat{\kappa} = \kappa \cdot (\cos(\theta), \sin(\theta)) \,, \, f > 0\right) \tag{16.28}$$

(— a superposition of solitons all having the *same* κ^2) of the Veselov–Novikov equations [7]

$$v_t = \partial^3 v + \bar{\partial}^3 v + \partial(wv) + \bar{\partial}(\bar{w}v) \quad ; \quad \begin{aligned} \partial &= \partial_1 - i\partial_2 \\ \bar{\partial} w &= -3\partial v \end{aligned} \quad \bar{\partial} = \partial_1 + i\partial_2 \qquad (16.29)$$

which are connected to the Manakov (L, A, B) triple

$$L = -\partial\bar{\partial} + v \quad , \quad A = \partial^3 + w\partial \quad , \quad B = (\partial w + \bar{\partial}\bar{w})$$
$$\bar{A} = \bar{\partial}^3 + \bar{w}\bar{\partial} \qquad (16.30)$$
$$\dot{L} = -[L, A + \bar{A}] + BL \quad .$$

Notes and references

The replacement of bilinear by multilinear operators is somewhat reminiscent of the tetrahedron– (or n–simplex) generalisations of the Yang Baxter equations. Poisson–brackets and area–preserving diffeomorphisms are vital to membrane theories and the $N \to \infty$ limit of $SU(N)$ Yang–Mills theories [8]. As underlying Lie–algebras for Lax formulations (see also [9],[10]) Poisson structures have been used in [11] as well as, related e.g. to self–dual gravity, in the very interesting paper [12].

[1]   R. Hirota; *Direct methods of finding solutions of nonlinear evolution equations* in Lect. Notes in Math. 515 (R.M. Miura, editor), Springer Verlag 1976.

[2]   M. Sato; *Soliton equations as dynamical systems on an infinite dimensional Grass-mannian manifold*, Kokyuroko, RIMS 439 (1981) p.30, Kyoto University.

[3]   see e.g. A.C. Newell, *Solitons in Math. and Phys.* CBMS–NFS Regional Conference Ser. in Appl. Math. 48, SIAM 1985

[4]   D.B. Fairlie, P. Fletcher, C. Zachos; Phys. Lett. B218 (89) 203.

[5]   J. Hoppe, M. Olshanetsky, S. Theisen; *Dynamical Systems on Quantum Tori Algebras*, KA–THEP–10/91

[6]   P.G. Grinevich, R.G. Novikov; Sov. Math. Dokl. #33(1986)#1 p.9.

[7]   S.P. Novikov, A.P. Veselov; Physica 18 D (1986) 267.

[8]   J. Goldstone; unpublished; J. Hoppe; MIT Ph.D. thesis 1982.
        J Hoppe; Part. Res. J. (Kyoto) 80,3 (1989/90).

[9]   M.I. Golenisheva–Kutuzova, A.G. Reiman; LOMI 169 (1988) 44.

[10]   M.V: Saveliev, A.M. Vershik; Comm. Math. Phys. 126 (1989) 367.

[11]   M. Bordemann, J. Hoppe, S. Theisen; Phys.Lett. 267B (1991) 374.

[12]   K. Takaski, T. Takebe; *SDiff(2) Toda* ... RIMS–790 (August 91).

This book was processed by the author using the TeX macro package from Springer-Verlag.